PHYSICS RESEARCH AND TECHNOLOGY

AF837471

ANALYTICAL MODELS OF THERMAL AND PHASE-TRANSFORMATION-INDUCED STRESSES IN MATERIALS WITH VOID DEFECTS I

PHYSICS RESEARCH AND TECHNOLOGY

Additional books in this series can be found on Nova's website under the Series tab.

Additional e-books in this series can be found on Nova's website under the e-book tab.

PHYSICS RESEARCH AND TECHNOLOGY

ANALYTICAL MODELS OF THERMAL AND PHASE-TRANSFORMATION-INDUCED STRESSES IN MATERIALS WITH VOID DEFECTS I

LADISLAV CENIGA

nova
science publishers
New York

Copyright © 2018 by Nova Science Publishers, Inc.

All rights reserved. No part of this book may be reproduced, stored in a retrieval system or transmitted in any form or by any means: electronic, electrostatic, magnetic, tape, mechanical photocopying, recording or otherwise without the written permission of the Publisher.

We have partnered with Copyright Clearance Center to make it easy for you to obtain permissions to reuse content from this publication. Simply navigate to this publication's page on Nova's website and locate the "Get Permission" button below the title description. This button is linked directly to the title's permission page on copyright.com. Alternatively, you can visit copyright.com and search by title, ISBN, or ISSN.

For further questions about using the service on copyright.com, please contact:
Copyright Clearance Center
Phone: +1-(978) 750-8400 Fax: +1-(978) 750-4470 E-mail: info@copyright.com.

NOTICE TO THE READER

The Publisher has taken reasonable care in the preparation of this book, but makes no expressed or implied warranty of any kind and assumes no responsibility for any errors or omissions. No liability is assumed for incidental or consequential damages in connection with or arising out of information contained in this book. The Publisher shall not be liable for any special, consequential, or exemplary damages resulting, in whole or in part, from the readers' use of, or reliance upon, this material. Any parts of this book based on government reports are so indicated and copyright is claimed for those parts to the extent applicable to compilations of such works.

Independent verification should be sought for any data, advice or recommendations contained in this book. In addition, no responsibility is assumed by the publisher for any injury and/or damage to persons or property arising from any methods, products, instructions, ideas or otherwise contained in this publication.

This publication is designed to provide accurate and authoritative information with regard to the subject matter covered herein. It is sold with the clear understanding that the Publisher is not engaged in rendering legal or any other professional services. If legal or any other expert assistance is required, the services of a competent person should be sought. FROM A DECLARATION OF PARTICIPANTS JOINTLY ADOPTED BY A COMMITTEE OF THE AMERICAN BAR ASSOCIATION AND A COMMITTEE OF PUBLISHERS.

Additional color graphics may be available in the e-book version of this book.

Library of Congress Cataloging-in-Publication Data

ISBN: 978-1-53613-526-8
ISSN: 2473-3679

Published by Nova Science Publishers, Inc. † New York

Táto kniha je venovaná s láskou mojim najdrahším rodičom.

This book is dedicated with love to my dearest parents.

Contents

Preface

This book[1] presents original analytical models of thermal and phase-transformation-induced stresses in *isotropic* components of two-component materials with void defects. These defects (i.e., pores, flaws) are a consequence of technological processes (e.g., powder metallurgy processes). These stresses, which are observed during a cooling process, originate below relaxation temperature of a two-component material. The thermal stresses are a consequence of different thermal expansion coefficients of material components. The phase-transformation-induced stresses are a consequence of different dimensions of crystalline lattices, which originate during a phase transformation. The void defects exhibit a significant influence on mechanical properties of materials, as well as on the thermal and phase-transformation-induced stresses. This influence is included within the analytical models in this book. Accordingly, this book can be considered to present unique analytical results.

With regard to the analytical models, real two-component materials with finite dimensions are replaced by a two-component model material system with infinite dimensions. This model material system is applicable to a real two-component material with precipitates in crystal grains. The precipitates and the crystal grains of this real material are considered to represent spherical particles and a matrix of the model material system, respectively. Similarly, this model material system is applicable to a real two-component material (e.g., dual-phase steel), which consists of two types of crystal grains (i.e., the type *A, B*) with different material properties. In this case, *A* and *B* are considered to represent spherical particles and a matrix of the model material system, respectively.

This multi-particle-matrix model system with isotropic spherical particles and spherical voids, both of which are periodically distributed in an isotropic infinite matrix, is imaginarily divided into cubic cells. Each cell contains a cen-

[1]This book was supported by the Slovak Grant Agency under the contract VEGA 2 / 0073 / 18.

tral particle or a central void. Accordingly, the cubic cell represents such a part of the model material system, which is related to one spherical particle and to one spherical void. The analytical models are based on this cell model. As presented in [1]–[4], cell models are used in case of analytical and computational models of phenomena in infinite periodic model material systems. As presented in [5], infinite matrixes are considered due to mathematical simplicity of analytical solutions. These solutions are assumed to exhibit sufficient accuracy due to the size of material components (e.g., particles, pores, flaws), which are relatively small in comparison with macroscopic material samples, macroscopic structural elements, etc.

The analytical models are determined by different mathematical procedures, which are applied to fundamental equations of solid continuum mechanics, i.e., Cauchy's equations, Hooke's law, the equilibrium equations. These different procedures result in different partial differential equations with non-zero right-hand sides.

These differential equations result in different mathematical solutions for the thermal and phase-transformation-induced stresses. These solutions are determined for each isotropic material component of the model material system. Additionally, this determination results from mathematical boundary conditions, which are defined for each isotropic material component. Finally, due to these different solutions, the principle of minimum total potential energy of an elastic solid body [6] is required to be considered. Strictly speaking, such solutions for the isotropic components are considered to exhibit minimum total potential energy of the model material system. The total potential energy W_t of the model material system is represented by the deformation energy W_d. Finally, W_d is represented by the elastic energy W_c, which is accumulated in the cubic cell, i.e., $W_t = W_d = W_c$.

Results of this book are applicable within basic research (solid continuum mechanics, theoretical physics, materials science), as well as within the practice of engineering.

With regard to the basic research, the analytical models in this book can be incorporated to the Eshelby's model [7]. The Eshelby's model and its development [7]–[9] define the disturbance of an applied stress-field in a solid continuum. The applied stress-field (e.g., mechanical loading) is disturbed due to the presence of inclusions in a solid continuum. This incorporation thus defines such a stress-strain state, which is a consequence of the thermal, phase-transformation-induced and mechanical stresses [5].

The analytical models in this book can be incorporated into analytical, computational and experimental models of material stresses [10]–[12], strengthening, interactions of energy barriers with dislocations and magnetic domain walls, etc.

As presented in [13]–[15], material scientists and material engineers determine a thermal-stress state in real two-component materials by the analytical model in [16], which is determined for an one-particle-matrix system. This model system with the particle volume fraction $v_p = 0$ and the void volume fraction $v_v = 0$ consists of an isotropic infinite matrix with an isotropic spherical particle. However, real two-components materials are characterized by $v_p > 0$ and $v_v > 0$.

In contrast to the analytical results in [16], this book[2] presents analytical models of the thermal and phase-transformation-induced stresses in the multi-particle-matrix system with void defects. Numerical values of these stresses can determined for different values of $v_p > 0$ and $v_v > 0$. Finally, in case of real two-component materials, this numerical determination for the multi-particle-matrix system with void defects is preferable in comparison with numerical results for the one-particle-matrix system [13]–[15].

[2] This book was reviewed by:

Dr. Pavel Diko, DrSc., Institute of Experimental Physics, Slovak Academy of Sciences, Košice, Slovak Republic.

Assoc. Prof. Marta Kianicová, PhD., Faculty of Industrial Technologies, Trenčín University of Alexander Dubček, Trenčín, Slovak Republic.

Assoc. Prof. Daniel Kottfer, PhD., Faculty of Mechanical Engineering, Technical University, Košice, Slovak Republic.

Chapter 1

Outline of Principles

As presented in Preface, this book presents analytical models of thermal and phase-transformation-induced stresses in isotropic components of two-component materials with void defects. These stresses, which are observed during a cooling process, originate below relaxation temperature of a two-component material (see Chapter 4). This analytical modelling includes:

- a definition of two-component materials, as well as a definition of such a model material system, which corresponds to these materials (see Chapter 2).
- a definition of such a coordinate system, which results from the model material system (see Sections 3.1, 3.2).
- an analysis of displacements of an arbitrary point of the model material system (see Section 3.3).
- definitions of Cauchy's equations and the equilibrium equations for this coordinate system. These equations and Hooke's law represent fundamental equations of solid continuum mechanics (see Sections 3.5–3.7).
- an analysis of mathematical boundary conditions for each isotropic component of the model material system (see Chapter 5).
- determinations of different mathematical procedures (see Sections 6.1, 7.1, 8.1), which are applied to these fundamental equations to obtain different mathematical solutions for these stresses. If different mathematical

solutions are determined, then the principle of minimum total potential energy [6] of the model system is required to be considered (see Section 3.9). This principle [6] means that an elastic solid body (i.e., the model system) tends to exhibit minimum total potential energy.

- mathematical determinations of final formulae for the thermal and phase-transformation-induced stresses in components of the model system (see Sections 6.2–6.6, 7.2–7.5, 8.2–8.5).
- selected topics in mathematics, which are required within the mathematical procedures (see Chapter 9).

Chapter 2 presents types of composite two-component materials with *isotropic* components, along with such a model system (i.e., a multi-particle-matrix system), which corresponds to these two-component materials. This multi-particle-matrix system is considered within the analytical modelling (see Section 2.2).

Numerical values of thermal stresses in two-component materials are usually determined by Selsing's analytical model [16]. This analytical model determines thermal stresses in a one-particle-matrix system with an isotropic spherical particle, which is embedded in an isotropic infinite matrix. On the one hand, parameters of this system, i.e., the inter-particle distance d and the particle volume fraction $v = V_p/V_m$, are not considered within Selsing's model.[1] On the other hand, this analytical model, which was determined in 1961, is still used by materials scientists at present [14, 15].

Mizutani's analytical model [17], which considers the inter-particle distance d and the particle volume fraction $v \in (0,1)$, determines thermal stresses in an infinite multi-particle-matrix system with isotropic components. In general, a thermal-stress state of multi-particle-matrix systems is determined by the stress $\sigma_{ij} = \sigma_{ji}$ [6] and the derivative $\partial\sigma_{ij}/\partial x_k$ $(i,j,k = 1,2,3)$, where x_i is an axis of the Cartesian system $(Ox_1x_2x_3)$. On the one hand, the stress σ_{ij} and the derivative $\partial\sigma_{ij}/\partial x_k$ $(i,j,k = 1,2,3)$ represent 24 variables, which are required to be determined by the fundamental equations of solid continuum mechanics (i.e., Cauchy's equations, Hooke's law, the equilibrium equations). This determination includes a mutual stress influence of particles. On the other hand, we get more variables than equations. Accordingly, these variables and this stress influence can not be determined.

[1] In case of Selsing's model, we get $v = V_p/V_m = 0$, $d \to \infty$, where V_p and $V_m \to \infty$ is volume of the particle and the infinite matrix, respectively.

As presented in [17], the infinite matrix is imaginarily divided into identical spherical cells with the diameter d, and the thermal stresses are determined within this cell. In case of this multi-particle-matrix system with isotropic components, the radial stress $\sigma_r = \sigma_{11}$, the tangential stress $\sigma_t = \sigma_{22} = \sigma_{33}$ and the derivatives $\partial\sigma_r/\partial r$, $\partial\sigma_t/\partial r$ can be determined by the fundamental equations, where r is one of the spherical coordinates (r, φ, ν), which are considered due to spherical particles. Due to the spherical cells, a mutual influence of cell surfaces is required to be determined by a suitable mathematical boundary condition for $r = d/2$.

As mentioned above, Mizutani's analytical model is based on spherical cells [17]. In this case, the matrix between three spherical cells is not considered. In contrast to the analytical results in [17], Section 2.1 defines such a multi-particle-matrix system, which is imaginarily divided into cubic cells with the dimension d (see Figure 2.1). This system consists of isotropic spherical particles and spherical voids. The particles and voids are periodically distributed in an isotropic infinite matrix. Each cell contains a central particle with the radius R_p or a central void with the radius R_v. Accordingly, the cubic cell represents such a part of the multi-particle-matrix system, which is related to one spherical particle with the volume fractions v_p and to one spherical void v_v. The thermal and phase-transformation-induced stresses are determined within this cubic cell, i.e., in the spherical particle, as well as in the cell matrix around the spherical particle and void. Accordingly, these stresses are functions of d, R_q, v_q, where the subscripts $q = p$ and $q = v$ are related to the spherical particle and void, respectively.

As presented in [1]–[4], analytical and computational models of phenomena in infinite periodic model systems are determined within identical suitable cells. Due to this infinity and periodicity, analytical and computational results, which are determined in a certain cubic cell, are valid for any cubic cell. Additionally, infinite matrixes are considered due to mathematical simplicity of analytical and computational solutions for material components (e.g., particles, pores, flaws), which are relatively small in comparison with macroscopic material samples, macroscopic structural elements, etc. Finally, in spite of this simplification, these solutions are acceptable [1]–[5].

Let the multi-particle-matrix system in Figure 2.1 have a finite matrix. Analytical models of the thermal and phase-transformation-induced stresses in this finite system are required to consider dimensions of the finite matrix, as well as a position of each cubic cell. Consequently, these stresses are required to be determined for each cubic cell. On the one hand, let final formulae for these stresses in this finite system be determined for each cubic cell. On the other hand, applications of these final formulae to real two-component materials with finite dimensions would be time-consuming.

Chapter 3 presents selected topics in solid continuum mechanics. In general, a stress-strain state of an elastic solid continuum is required to be determined at each point of this continuum, i.e., at an arbitrary point. The arbitrary point is determined by a suitable coordinate system. Consequently, an infinitesimal part of this continuum is considered at this arbitrary point. An influence of the continuum on the infinitesimal part is represented by displacements and stresses [6]. The stresses act on this infinitesimal surface. The displacements of the infinitesimal part, along with the stresses, are investigated along axes of the coordinate system.

Due to spherical particles and voids of the multi-particle-matrix model system in Figure 2.1, the arbitrary point P is determined by the spherical coordinates (r, φ, ν) (see Section 3.1), which are related to the Cartesian system $(Ox_1x_2x_3)$ (see Figure 3.1a), where O is a centre of the spherical particle and void (see Figure 2.1). Consequently, an infinitesimal spherical cap is considered (see Figure 3.1b). The displacements and stresses are investigated along the axes x_1', x_2', x_3' of the Cartesian system $(Px_1'x_2'x_3')$, where $r = |OP|$. The axes $x_1' = x_r$ and $x_2' = x_\varphi$, $x_3' = x_\nu$ at the point P represent radial and tangential directions (see Figure 3.1a), respectively. The conventional subscripts r, φ, ν, which are used in a connection with the spherical coordinates (r, φ, ν), are replaced by the subscripts 1, 2, 3 (see Section 3.4), respectively.

The infinite multi-particle-matrix system in Figure 2.1 is symmetric. Due to this infinity and symmetry, the thermal and phase-transformation-induced stresses are sufficient to be investigated within one eighth of the cubic cell (see Figure 3.2), i.e., for $\varphi \in \langle 0, \pi/2 \rangle$ and $\nu \in \langle 0, \pi/2 \rangle$. Intervals of the variable r for components of the model system are determined in Section 3.2. The displacements of the infinitesimal spherical cap at the arbitrary point P are analysed in Section 3.3. This cap exhibits a radial displacement along the axis $x_1' = x_r$.

The analytical models of the thermal and phase-transformation-induced stresses are determined by different mathematical procedures. These procedures are applied to a system of differential equations (see Section 3.8). This system is determined by fundamental equations of solid continuum mechanics, i.e., Cauchy's equations, Hooke's law, the equilibrium equations (see Sections 3.5–3.7). These applications result in different partial differential equations with non-zero right-hand sides. Due to these different equations, the thermal and phase-transformation-induced stresses are determined by different solutions for isotropic components of the multi-particle-matrix system. This determination results from mathematical boundary conditions, which are defined for each component.

Due to these different solutions, the principle of minimum total potential energy of an elastic solid body [6] is required to be considered. Strictly speaking, such solutions for the isotropic components are considered to exhibit minimum total potential energy of the multi-particle-matrix system. As analysed in Section 3.9, the total potential energy W_t is represented by the deformation energy W_d. Finally, W_d is represented by the elastic energy W_c, which is accumulated in the cubic cells, i.e., $W_t = W_d = W_c$.

Chapter 4 The thermal and phase-transformation-induced stresses in two-component materials originate during a cooling process at the temperature $T < T_r$, where the relaxation temperature T_r is defined in Section 4.1. If $T \geq T_r$, then these stresses are relaxed by thermal-activated processes [18]. The thermal stresses are a consequence of different thermal expansion coefficients of material components (see Section 4.2). The phase-transformation-induced stresses are a consequence of different dimensions of crystalline lattices, which are transformed at the phase-transformation temperature $T_{tq} \leq T_r$, where the subscripts $q = p$ and $q = m$ are related to the spherical particle and matrix, respectively. These different dimensions result in the phase-transformation-induced strain ε_{tq} (see Section 4.3), which is a reason of the phase-transformation-induced stresses.

Chapter 5 The different solutions for the thermal and phase-transformation-induced stresses in isotropic components of the multi-particle-matrix system are determined by mathematical mandatory and additional boundary conditions. The mandatory conditions (see Sections 5.1–5.3) are determined for the spherical particle, as well as for the cell matrix around the spherical particle and void. The additional condition is determined for the cell matrix around the spherical particle and void (see Section 5.4). This condition is based on the following analysis.

The identical cubic cells of the multi-particle-matrix system in Figure 2.1 are imaginary, i.e., their surface is not a physical boundary. Let A and B represent neighbouring cubic cells. Let the identical functions $u'_{1mA} = u'_{1mA}(r, \varphi, \nu)$, $u'_{1mB} = u'_{1mB}(r, \varphi, \nu)$ represent radial displacements in the cell matrix of these neighbouring cells. Let C be an arbitrary point on a common surface of these cells (see Figure 5.1). Let the identical functions $(u'_{1mA})_C$, $(u'_{1mB})_C$ are connected at the point C, where $(u'_{1mA})_C = (u'_{1mB})_C$. This connection is assumed to be 'smooth'. Finally, the additional condition results from this mathematical assumption.

Chapters 6, 7, 8 present analytical models of the thermal and phase-transformation-induced stresses. These models are determined by different mathematical procedures (see Sections 6.1, 7.1, 8.1), which are applied to a system of differential equations (see Section 3.8). This system is determined by fundamental equations of solid continuum mechanics (see Sections 3.5–3.7). These procedures result in different partial differential equations with non-zero right-hand sides. Due to these different equations, the thermal and phase-transformation-induced stresses are determined by different solutions for isotropic components of the multi-particle-matrix system. This determination results from mathematical boundary conditions, which are defined for each component. Final formulae for these stresses are determined in Sections 6.2–6.6, 7.2–7.5, 8.2–8.5.

Chapter 9 presents such a part of mathematics, which is required within mathematical procedures in this book.

Chapter 2

Multi-Particle-Matrix System

2.1. Model Material System

As presented in Preface, this book deals with analytical models of the thermal stresses and phase-transformation-induced stresses in isotropic components of two-component materials with void defects. These defects (i.e., pores, flaws) are a consequence of technological processes (e.g., powder metallurgy processes) [19]–[24]. The two-component materials with finite dimensions consist of

1. precipitates, which are distributed in crystal grains,
2. two types of crystal grains (i.e., type A, B) with different material properties.

The analytical models are determined for such a two-component model material system with infinite dimensions, which corresponds to these two types of two-component materials. As presented in Figure 2.1a, this multi-particle-matrix model system, which consists of isotropic spherical particles, spherical voids and an isotropic infinite matrix, is imaginarily divided into cubic cells of the type *I* and *II*, where d is a dimension of these cells. The spherical particles with the radius R_p and the spherical voids with the radius R_v are periodically distributed in this matrix. Each cell contains a central spherical particle (*I*) or a central spherical void (*II*). Accordingly, the cubic cell represents such a part of the model material system, which is related to one spherical particle and one spherical void. The matrix is infinite along the axes x_1, x_2, x_3 of the Cartesian system $(Ox_1x_2x_3)$, where O is identical with a centre of the spherical particles and voids. The model material system in Figure 2.1a is depicted in the plane x_1x_2 of the Cartesian system $(Ox_1x_2x_3)$, where O is identical with a centre of the cubic cell. Due to infinity of the matrix, the same figure is also considered for the planes x_1x_3 and x_2x_3.

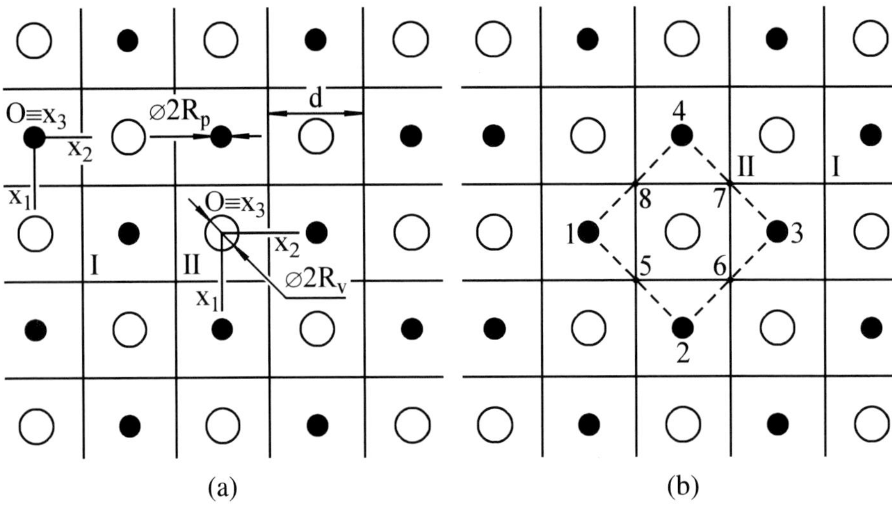

Figure 2.1. (a) The multi-particle-matrix system is imaginarily divided into cubic cells of the type *I* and *II* with a central spherical particle and a central spherical void with the radius R_p and R_v, respectively, where d is a dimension of these cells. The matrix is infinite along the axes x_1, x_2, x_3 of the Cartesian system $(Ox_1x_2x_3)$, where O is identical with a centre of the spherical particles and voids. (b) The cell *1234* with the dimensions $\sqrt{2}d \times \sqrt{2}d \times d$ is related to the central cubic cell *5678* of the type *II*, where *1234* represents a definition cell of the multi-particle-matrix system. The quarters *158*, *265*, *376*, *487* represent a cubic cell of the type *I*, i.e., *I=158+265+376+487*, and then we get *1234=I+II*.

As presented in Figure 2.1b, the multi-particle-matrix system is defined by the cell *1234*. This definition cell with the dimensions $\sqrt{2}d \times \sqrt{2}d \times d$ and with the volume $V_{1234} = 2d^3$ is related to the central cubic cell *5678* of the type *II*. The quarters *158*, *265*, *376*, *487* represent a cubic cell of the type *I*, i.e., *I=158+265+376+487*. The cell *1234* consists of one cubic cell of the type *I* and one cubic cell of the type *II* (see Figure 2.1a), i.e., *1234=I+II*.

With regard to $V_q = 4\pi R_q^3/3$ ($q = p, v$), $V_{1234} = 2d^3$, the volume fraction $v_q = V_q/V_{1234}$ is derived as

$$v_q = \frac{2\pi}{3}\left(\frac{R_q}{d}\right)^3 \in \left(0, \frac{\pi}{12}\right\rangle, \quad q = p, v, \tag{2.1}$$

where the subscripts $q = p$ and $q = m$ are related to the spherical particle and void, respectively. The maximum value $v_{q\,max} = \pi/12$ is determined on the condition $d = 2R_q$. With regard to Equation (2.1) for $q = p$, we get

$$d = R_p \left(\frac{2\pi}{3v_p}\right)^{1/3}, \tag{2.2}$$

where d, R_q, v_q represent parameters of the multi-particle-matrix system. In case of a real two-component material, these parameters are determined by computational techniques, which are applied to experimental results. This determination is related to the temperature $T \in \langle T_f, T_r \rangle$, where T_f is final temperature of a cooling process, and the relaxation temperature T_r is analysed in Section 4.1.

As presented in [1]–[5], analytical and computational models of phenomena in infinite periodic model systems are determined within identical suitable cells. Due to this infinity and periodicity, analytical and computational results, which are determined in a certain cubic cell, are valid for any cubic cell. With regard to the infinite multi-particle-matrix system in Figure 2.1, the thermal and phase-transformation-induced stresses are determined for the spherical particle, as well as for the cell matrix around the spherical particle and the spherical void. Formulae for these stresses, which are determined for a certain cubic cell of the type N ($N = I, II$), are valid for each of these cells in the infinite matrix.

Additionally, infinite matrixes are considered due to mathematical simplicity of analytical and computational solutions for material components (e.g., particles, pores, flaws), which are relatively small in comparison with macroscopic material samples, macroscopic structural elements, etc. Finally, in spite of this simplification, these solutions are acceptable [1]–[5].

Let the multi-particle-matrix system in Figure 2.1 have a finite matrix. Analytical models of the thermal and phase-transformation-induced stresses in this finite system are required to consider dimensions of the finite matrix, as well as a position of each cubic cell. Consequently, these stresses are required to be determined for each cubic cell. On the one hand, let final formulae for these stresses in this finite system be determined for each cubic cell. On the other hand, applications of these final formulae to real two-component materials with finite dimensions would be time-consuming.

2.2. Components of Model Material System

As presented above (see Item 1), the precipitates and the crystal grains are considered to represent the spherical particles and the matrix of the multi-particle-matrix system, respectively. Similarly, let the crystal grains A and B (see Item 2 be characterized by the volume fractions $v_A \in \langle 0, v_{AB} \rangle$ and $v_B \in \langle 0, v_{AB} \rangle$, respectively, where $v_A + v_B = v_{AB}$.[1] If $v_A < v_B$, then the grains A and B are considered to represent the spherical particles and the matrix of the multi-particle-matrix system, respectively. If $v_A > v_B$, then the grains A and B are considered to represent the matrix and the spherical particles of the multi-particle-matrix system, respectively.

If $v_A = v_B = v_{AB}/2$, then the following analysis is required to be considered. As presented in Figure 2.1b, the multi-particle-matrix system consists of the cells *1234*. This analysis results from the elastic energy W_c, which is accumulated in the cell *1234*.

Let the elastic energy W_{cAB} (see Equation (3.41)) is related to such a multi-particle-matrix system, when the crystal grains A and B are considered to represent the spherical particles and the matrix, respectively.

Let the elastic energy W_{cBA} (see Equation (3.41)) is related to such a multi-particle-matrix system, when the crystal grains B and A are considered to represent the spherical particles and the matrix, respectively.

Let W_{cAB} and W_{cBA} be numerically determined for a two-component material, which is defined in Section 2.1 (see Item 2, page 7). If $W_{cAB} < W_{cBA}$, then the crystal grains A and B of this two-component material are considered to represent the spherical particles and the cell matrix, respectively. If $W_{cAB} > W_{cBA}$, then the crystal grains B and A of this two-component material are considered to represent the spherical particles and the cell matrix, respectively.

---

[1] In case of the multi-particle-matrix system in Figure 2.1, we get $v_{AB} = 1 - v_v$.

Chapter 3

Solid Continuum Mechanics

As usual in solid continuum mechanics, stresses in a solid continuum are investigated within an infinitesimal part of the solid continuum at an arbitrary point. A position of this infinitesimal part is determined by a suitable system of coordinates. A shape of this part results from that of the solid continuum.

An influence of the solid continuum on this part is represented by displacements of this part, as well as by stresses, which act on surfaces of this part [6]. Relationships between these displacement and strains of this part are defined by Cauchy's equations (see Section 3.5). Similarly, relationships between these stresses are defined by the equilibrium equations (see Section 3.6). Finally, relationships between these strains and stresses are defined by Hooke's law (see Section 3.7). Cauchy's equations, the equilibrium equations, along with Hooke's law, represent fundamental equations of solid continuum mechanics.

If a system of these fundamental equations exhibits different solutions, then the principle of minimum total potential energy [6] is required to be considered (see Section 3.8). Strictly speaking, such a solution is considered, which exhibits minimum total potential energy of the material model system (see Figure 2.1). Finally, such a stress-strain state of the model material system is realized to correspond to this solution.

Chapter 3 deals with

- a definition of such a coordinate system and an infinitesimal part, which are suitable for the model material system (see Section 3.1),
- a definition of intervals of coordinates within the cubic cells *I*, *II* (see Figure 2.1, Section 3.2),

- an analysis of the displacements in these cubic cells (see Section 3.3),
- a definition of the fundamental equations for the infinitesimal part (see Section 3.4),
- a definition of elastic energy of the model material system with respect to the principle of minimum total potential energy (see Section 3.5) [6],
- a definition of the mathematical condition to determine a radial stress, which acts at the particle-matrix boundary of the model material system (see Section 3.6),
- an analysis of a temperature dependence of the thermal stresses and phase-transformation-induced stresses (see Section 3.7).

3.1. Coordinate System

As presented in Figure 3.1a, the thermal and phase-transformation-induced stresses are determined at the arbitrary point P along the axes x_1', x_2', x_3' of the Cartesian system $(Px_1'x_2'x_3')$, where P is defined by the spherical coordinates (r, φ, ν).

The axes $x_1' = x_r$ and $x_2' = x_\varphi$, $x_3' = x_\nu$ at the point P represent radial and tangential directions related to the spherical surface with the radius $r = |OP|$, respectively, where $x_\varphi \parallel x_1 x_2$.[1] Finally, the infinitesimal part at the point P is represented by an infinitesimal spherical cap with the parameters dr, $d\varphi$, $d\nu$ (see Figures 3.3, 3.4), where dr is a dimension along the axis x_1'.

As presented in Figure 3.1b, the surface $S_r = A_1B_1C_1D_1$ of the infinitesimal spherical cap at the radius $r = \left|\overline{OP}\right|$ has the dimensions $|A_1D_1| = |B_1C_1| = r \times d\varphi$ and $|A_1B_1| = |C_1D_1| = r \times d\nu$ along the axes x_2' and x_3', respectively.

Similarly, the surface $S_{r+dr} = A_2B_2C_2D_2$ of the infinitesimal spherical cap at the radius $r + dr$ has the dimensions $|A_2D_2| = |B_2C_2| = (r + dr) \times d\varphi$ and $|A_2B_2| = |C_2D_2| = (r + dr) \times d\nu$ along the axes x_2' and x_3', respectively[2].

[1]The conventional subscripts r, φ, ν, which are used in connection with the spherical coordinates (r, φ, ν), are replaced by the subscripts 1, 2, 3, respectively, i.e., $x_1' = x_r$, $x_2' = x_\varphi$, $x_3' = x_\nu$ (see Figure 3.1a).

[2]The 'conventional approach' [6] considers the dimensions $|A_1D_1| = |B_1C_1| = r \sin\nu \, d\varphi$ and $|A_2D_2| = |B_2C_2| = (r + dr) \sin\nu \, d\varphi$ along the axis x_2' (see Figure 3.1) at the radii r and $r + dr$, respectively.

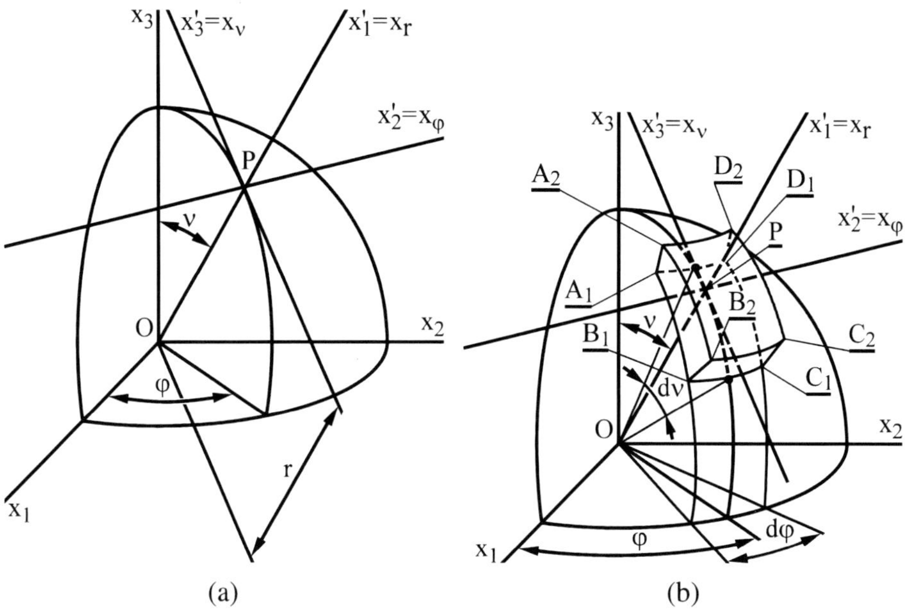

Figure 3.1. (a) The arbitrary point P defined by the spherical coordinates (r, φ, ν) with respect to the Cartesian system $(Ox_1x_2x_3)$ (see Figure 2.1a). The axes $x_1' = x_r$ and $x_2' = x_\varphi$, $x_3' = x_\nu$ represent radial and tangential directions, respectively, where $x_\varphi \parallel x_1x_2$. (b) The infinitesimal spherical cap at the point P with the surfaces $S_r = A_1B_1C_1D_1$ and $S_{r+dr} = A_2B_2C_2D_2$ at the radii $r = |\overline{OP}|$ and $r + dr$, respectively, where $|A_1A_2| = |B_1B_2| = |C_1C_2| = |D_1D_2| = dr$, $|A_1D_1| = |B_1C_1| = r \times d\varphi$, $|A_1B_1| = |C_1D_1| = r \times d\nu$, $|A_2D_2| = |B_2C_2| = (r + dr) \times d\varphi$, $|A_2B_2| = |C_2D_2| = (r + dr) \times d\nu$.

The axis x_1' represents a normal of S_r and S_{r+dr}, where $|A_1A_2| = |B_1B_2| = |C_1C_2| = |D_1D_2| = dr$.

The spherical coordinates (r, φ, ν) and the infinitesimal spherical cap are considered due to the spherical particles and voids (see Figure 2.1).

3.2. Intervals of Coordinates

As presented in Section 2.2, the thermal and phase-transformation-induced stresses are determined within the cubic cells *I*, *II* (see Figure 2.1). The multi-particle-matrix system in Figure 2.1, is symmetric. This symmetry results from infinity of the matrix, as well as from periodical distributions of the spherical particles and voids. Accordingly, as presented in Figure 3.2, these stresses are sufficient to be determined within one eighth of the cubic cell N ($N = I, II$), i.e., for $\varphi \in \langle 0, \pi/2 \rangle$ and $\nu \in \langle 0, \pi/2 \rangle$.

The interval $r \in \langle 0, R_p \rangle$ is related to the spherical particle (see Figure 2.1a). The intervals $r \in \langle R_p, r_c \rangle$ and $r \in \langle R_v, r_c \rangle$ are related to the matrix of the cubic cell *I* and *II*, respectively. Additionally, the distance r_c is considered within the mathematical boundary conditions, which are defined in Chapter (see Equations (5.1)–(5.8)).

As presented in Figure 3.2, we get $r_c = OC_1$ and $r_c = OC_2$ for $\nu \in \langle 0, \nu^* \rangle$ and $\nu \in \langle \nu^*, \pi/2 \rangle$, respectively. The points C_1 and C_2 represent intersections of the axis x_1' with the surfaces *3657* and *1456*, respectively. The axis x_1' represents a radial direction (see Figure 3.1a), which is defined by the angles $\varphi \in \langle 0, \pi/2 \rangle$ and $\nu \in \langle 0, \pi/2 \rangle$. The angle ν^* (see Figure 3.2) is derived as

$$\nu^* = \arctan\left(\frac{|\overline{39}|}{|\overline{O3}|}\right) = \arctan\left(\frac{1}{c_\varphi}\right), \tag{3.1}$$

where $|\overline{O3}| = d/2$ and $|\overline{39}| = 1/(2c_\varphi)$ represent length of the abscissae $\overline{O3}$ and $\overline{39}$, respectively. The coefficient c_φ has the forms

$$c_\varphi = \cos\varphi, \quad \varphi \in \left\langle 0, \frac{\pi}{4} \right\rangle, \tag{3.2}$$

$$c_\varphi = \sin\varphi, \quad \varphi \in \left\langle \frac{\pi}{4}, \frac{\pi}{2} \right\rangle. \tag{3.3}$$

In case of the interval $\nu \in \langle 0, \nu^* \rangle$ (see Figure 3.2), the distance $r_c = |\overline{OC_1}|$ ($N = I, II$) is determined by the following mathematical procedures. Considering the sine rule [25]

$$\frac{|\overline{C_1 9}|}{\sin\left[\angle\left(\overline{OC_1}, \overline{O9}\right)\right]} = \frac{|\overline{O9}|}{\sin\left[\angle\left(\overline{OC_1}, \overline{C_1 9}\right)\right]} \tag{3.4}$$

and the condition

$$|\overline{O9}| = r_c \cos\left[\angle\left(\overline{OC_1}, \overline{O9}\right)\right] + |\overline{C_1 9}| \cos\left[\angle\left(\overline{C_1 9}, \overline{O9}\right)\right], \tag{3.5}$$

we get

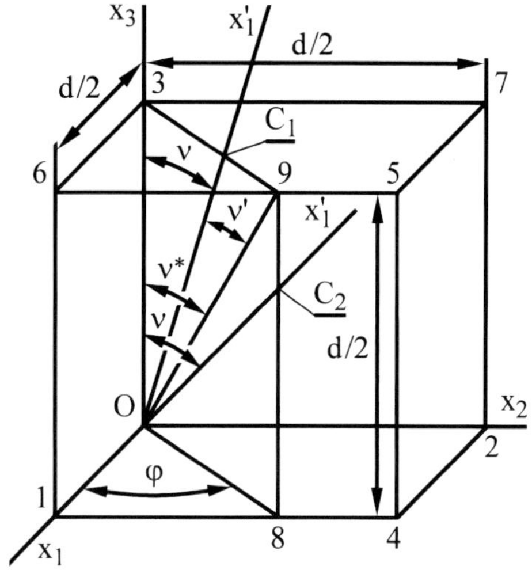

Figure 3.2. One eighth of the cubic cell N ($N = I, II$) (see Figure 2.1). The points C_1 and C_2 represent intersections of the axis x_1' with the surfaces *3657* and *1456* for for $\nu \in \langle 0, \nu^* \rangle$ and $\nu \in \langle \nu^*, \pi/2 \rangle$ (see Equations (3.1)–(3.3)), respectively. The axis x_1' represents a radial direction (see Figure 3.1a), which is defined by the angles $\varphi \in \langle 0, \pi/2 \rangle$ and $\nu \in \langle 0, \pi/2 \rangle$.

$$r_c = \frac{\left|\overline{O9}\right|}{\cos\left[\angle\left(\overline{OC_1}, \overline{O9}\right)\right]} \left\{ 1 - \frac{\sin\left[\angle\left(\overline{OC_1}, \overline{O9}\right)\right] \cos\left[\angle\left(\overline{C_19}, \overline{O9}\right)\right]}{\sin\left[\angle\left(\overline{OC_1}, \overline{C_19}\right)\right]} \right\}. \tag{3.6}$$

With regard to $\left|\overline{O3}\right| = d/2$, $\left|\overline{39}\right| = 1/\left(2c_\varphi\right)$, the length $\left|\overline{O9}\right|$ of the abscissa $\overline{O9}$ and the angles $\angle\left(\overline{OC_1}, \overline{O9}\right)$, $\angle\left(\overline{C_19}, \overline{O9}\right)$, $\angle\left(\overline{OC_1}, \overline{C_19}\right)$ are derived as

$$\left|\overline{O9}\right| = \sqrt{\left|\overline{O3}\right|^2 + \left|\overline{39}\right|^2} = \frac{d \times \sqrt{1 + c_\varphi^2}}{2c_\varphi}, \tag{3.7}$$

$$\angle\left(\overline{OC_1}, \overline{O9}\right) = \nu' = \nu^* - \nu, \quad \nu' \in \langle 0, \nu^* \rangle, \tag{3.8}$$

$$\angle\left(\overline{C_19}, \overline{O9}\right) = \frac{\pi}{2} - \nu^*, \tag{3.9}$$

$$\angle\left(\overline{OC_1},\overline{C_19}\right)=\frac{\pi}{2}+\nu^*-\nu'. \qquad (3.10)$$

Considering the following relationships [19]

$$\sin\left(\frac{\pi}{2}\pm\alpha\right)=\cos\alpha, \qquad (3.11)$$

$$\cos\left(\frac{\pi}{2}\pm\alpha\right)=\mp\sin\alpha, \qquad (3.12)$$

$$\sin(\alpha\pm\beta)=\sin\alpha\cos\beta\pm\cos\alpha\sin\beta, \qquad (3.13)$$

$$\cos(\alpha\pm\beta)=\cos\alpha\cos\beta\mp\sin\alpha\sin\beta, \qquad (3.14)$$

the distance $r_c=\left|\overline{OC_1}\right|$ (see Figure 3.1) has the form

$$r_c=\left|\overline{OC_1}\right|=d\,f_c. \qquad (3.15)$$

The function $f_c=f_c(\varphi,\nu')$ is derived as

$$f_c=\frac{\sqrt{1+c_\varphi^2}}{2\left(\sin\nu'+c_\varphi\cos\nu'\right)}, \quad \nu'\in\langle 0,\nu^*\rangle, \qquad (3.16)$$

where the coefficient $c_\varphi=c_\varphi(\varphi)$ is given by Equations (3.2), (3.3).

In case of the interval $\nu\in\langle\nu^*,\pi/2\rangle$ (see Figure 3.2), the distance $r_c=\left|\overline{OC_2}\right|$ is derived as

$$r_c=\left|\overline{OC_2}\right|=\frac{d}{2c_\varphi\sin\nu}, \qquad (3.17)$$

The function $f_c=f_c(\varphi,\nu)$ has the form

$$f_c=\frac{1}{2c_\varphi\sin\nu}, \qquad (3.18)$$

where the coefficient $c_\varphi=c_\varphi(\varphi)$ is given by Equations (3.2), (3.3).

3.3. Radial Displacement

The thermal and phase-transformation-induced stresses are a consequence of the condition $\beta_p \neq \beta_m$, where the coefficient β_q $(q = p,m)$ is presented in Section 4.1 (see Equations (4.1)–(4.7)). The subscript $q = p$ and $q = m$ is related to the spherical particle and matrix, respectively. These stresses originate at the temperature $T \in \langle T_f, T_r \rangle$, where T_f is final temperature of a cooling process , and the relaxation temperature T_r is analysed in Section 4.1.

If $T \in \langle T_f, T_r \rangle$ and $\beta_p < \beta_m$, then the matrix of the cubic cell *I* is pushed by the spherical particle. Consequently, the spherical particle is pushed by the matrix of the cubic cell *I*. If $T \in \langle T_f, T_r \rangle$ and $\beta_p > \beta_m$, then the matrix of the cubic cell *I* is pulled by the spherical particle. Consequently, the spherical particle is pulled by the matrix of the cubic cell *I*. The following analysis of this pulling (or pushing) is required to be considered.

As presented in Section 3.2, the multi-particle-matrix system is symmetric. Due to this symmetry, the pulling (or pushing) of an arbitrary point at the particle-matrix boundary in the cubic cell *I* is realized along a normal to this boundary, i.e., along the axis x_1', which represents a radial direction (see Figures 2.1, 3.1a). Let *P* be an arbitrary point at the particle-matrix boundary, where (R_p, φ, ν) are spherical coordinates of this point (see Figures 2.1, 3.1a). The infinitesimal spherical cap at the point *P* (see Figure 3.1b) exhibits a displacement along the axis x_1', i.e., the radial displacement $u_1' = u_r$. The radial displacement is a consequence of the radial stress $p_p = p_p(\varphi, \nu)$, which acts along the axis x_1' at the particle-matrix boundary, where is determined by Equation (3.45). Due to this radial stress, each point of the axis x_1' exhibits the radial displacement $u_{1q}' = u_{1q}'(r, \varphi, \nu)$ $(q = p,m)$ for $r \in \langle 0, r_c \rangle$, $\varphi \in \langle 0, \pi/2 \rangle$, $\nu \in \langle 0, \pi/2 \rangle$ (see Equations (3.1)–(3.3), (3.8), (3.15), (3.16)).[3] The subscript $q = p$ is related to the spherical particle, where $r \in \langle 0, R_p \rangle$ (see Figure 2.1a). The subscript $q = m$ is related to the matrix of the cubic cell *I* (see Figure 2.1), where $r \in \langle R_p, r_c \rangle$.

With regard to the boundary condition (5.4), the thermal and phase-transformation-induced stresses in the cubic cell *I* (see Figure 2.1) with the di-

[3]Let the condition $\beta_p = \beta_p$ be valid. If the temperature T increases (or decreases) within the interval $\langle T_f, T_r \rangle$, then the spherical particle and the matrix expand (or contract). This expansion (or contraction) is a reason of a radial displacement of each point of the axis x_1'. With regard to $\beta_p = \beta_p$, this radial displacement is a consequence of a temperature change within the interval $\langle T_f, T_r \rangle$. Due to $\beta_p = \beta_p$, the thermal and phase-transformation-induced stresses do not originate in the multi-particle-matrix system.

mension d result in the transformation $d \rightarrow d + u_{1c)}$. With regard to the boundary condition (5.6), these stresses result in the transformation $d \rightarrow d - u_{1c)}$, which is related to the cubic cell *II*. Finally, each point of the axis x_1' in the cubic cell *II* exhibits the radial displacement $u_{1v}' = u_{1v}'(r, \varphi, \nu)$, where $r \in \langle R_v, r_c \rangle$, $\varphi \in \langle 0, \pi/2 \rangle$, $\nu \in \langle 0, \pi/2 \rangle$.

3.4. Notation of Subscripts

As presented in Section 3.1, the infinitesimal spherical cap and the spherical coordinates (r, φ, ν) are considered due to the spherical particles and voids (see Figures 2.1, 3.1).

The thermal and phase-transformation-induced stresses are determined within the Cartesian system $(Px_1'x_2'x_3')$. With regard to the Cartesian system $(Ox_1x_2x_3)$ (see Figure 3.1), the point P is determined by $r = |OP|$, $\varphi \in \langle 0, \pi, 2 \rangle$, $\nu \in \langle 0, \pi, 2 \rangle$.

As presented in Figure 3.1, the conventional notation x_r, x_φ, x_ν is replaced by x_1', x_2', x_3', respectively. Consequently, with regard to $x_1' = x_r$, $x_2' = x_\varphi$, $x_3' = x_\nu$, the conventional notation for

- the radial displacement u_r is replaced by u_1'.
- the radial stress σ_r and the radial strain ε_r are replaced by σ_{11}' and ε_{11}', respectively.
- the tangential stresses σ_φ, σ_ν and the tangential strains ε_φ, ε_ν are replaced by σ_{22}', σ_{33}' and ε_{22}', ε_{33}', respectively.
- the shear stresses $\sigma_{r\varphi}$, $\sigma_{r\nu}$, $\sigma_{\varphi\nu}$ and the shear strains $\varepsilon_{r\varphi}$, $\varepsilon_{r\nu}$, $\varepsilon_{\varphi\nu}$ are replaced by σ_{12}', σ_{13}', σ_{23}' and ε_{12}', ε_{13}', ε_{23}', respectively.

Additionally, the reason of the notation ε_{ii}' and σ_{ii}' instead of ε_i' and σ_i' $(i = 1,2,3)$ is as follows, respectively.

The strain ε_{ij}' $(i, j = 1,2,3)$ is related to such a surface of the infinitesimal cap, which is defined by the normal x_j'. The strain ε_{ij}' is realized along the axis x_i', where $\varepsilon_{ij}' = \varepsilon_{ji}'$ [6]. As presented in Figure 3.1a, the axes x_1' and x_2', x_3' define radial and tangential directions, respectively. Consequently, ε_{11}' and ε_{22}', ε_{33}' and ε_{12}', ε_{13}' represent radial and tangential and shear strains, respectively.

The stress σ'_{ij} $(i, j = 1,2,3)$ acts on such a surface of the infinitesimal cap, which is defined by the normal x'_j. The stress σ'_{ij} $(i, j = 1,2,3)$ act along the axis x'_i, where $\sigma'_{ij} = \sigma'_{ji}$ [6]. Consequently, σ'_{11} and σ'_{22}, σ'_{33} and σ'_{12}, σ'_{13} represent radial and tangential and shear stresses, respectively.

3.5. Cauchy's Equations

Cauchy's equations define relationships between displacements and strains of an infinitesimal part of a solid continuum. In case of the infinitesimal spherical cap in the arbitrary point P (see Figure 3.1), the radial strain ε'_{11} (see Figure 3.3a,b), the tangential strains ε'_{22}, ε'_{33} (see Figure 3.3b), along with the shear strains ε'_{12}, ε_{13} (see Figure 3.4), are derived as

$$\varepsilon'_{11} = \frac{|1'''1'| - |1''1|}{|1''1|} = \frac{1}{dr}\left[\left(dr + \frac{\partial u'_1}{\partial r} dr\right) - dr\right] = \frac{\partial u'_1}{\partial r}, \tag{3.19}$$

$$\varepsilon'_{22} = \frac{|1'2'| - |12|}{|12|} = \frac{(u'_1 + r)\, d\varphi - r\, d\varphi}{r\, d\varphi} = \frac{u'_1}{r}, \tag{3.20}$$

$$\varepsilon'_{33} = \frac{|1'3'| - |13|}{|13|} = \frac{(u'_1 + r)\, d\nu - r\, d\nu}{r\, d\nu} = \frac{u'_1}{r}, \tag{3.21}$$

$$\varepsilon'_{12} = \tan\left[\angle\left(|12|, |12^*|\right)\right] = \frac{1}{r\, d\varphi}\left(\frac{\partial u'_1}{\partial \varphi} d\varphi\right) = \frac{1}{r}\frac{\partial u'_1}{\partial \varphi}, \tag{3.22}$$

$$\varepsilon'_{13} = \tan\left[\angle\left(|13|, |13^*|\right)\right] = \frac{1}{r\, d\nu}\left(\frac{\partial u'_1}{\partial \nu} d\nu\right) = \frac{1}{r}\frac{\partial u'_1}{\partial \nu}. \tag{3.23}$$

As presented in Section 3.3, the infinitesimal spherical cap exhibits the radial displacement u'_1 along the axis x'_1 (see Figure 3.1). Accordingly, we get $u'_2 = u_\varphi = 0$, $u'_3 = u_\nu = 0$, where u'_2 and u'_2 are tangential displacements along the axes x'_2 and x'_3, respectively. Finally, we get $\varepsilon'_{23} = \varepsilon'_{\varphi\nu} \propto \left[\left(\partial u'_2/\partial \nu\right) + \left(\partial u'_3/\partial \varphi\right)\right] = 0$, where ε'_{23} is a shear strain.

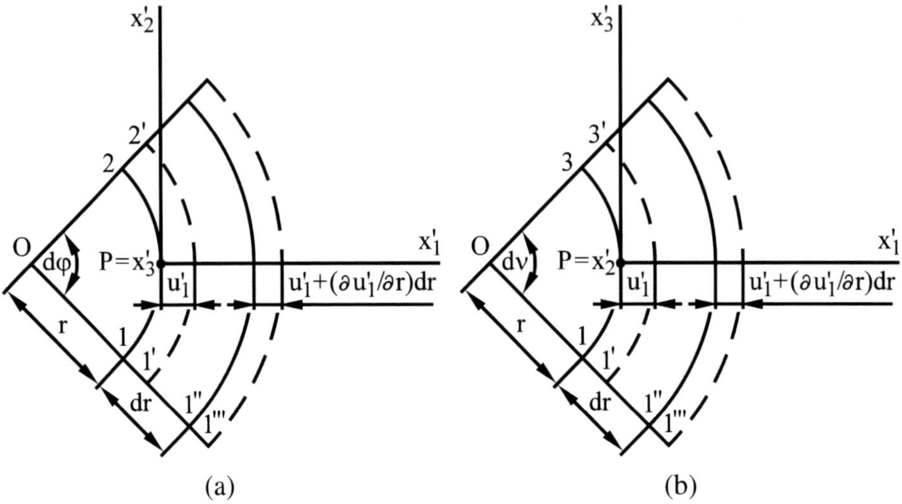

Figure 3.3. The radial displacements u_1' and $u_1' + (\partial u_1'/\partial r)\,dr$ of the infinitesimal spherical cap at the point P (see Figure 3.1) at the radii r and $r+dr$ in the planes (a) $x_1'x_2'$, (b) $x_1'x_3'$ (see Figure 3.1), respectively. In this case, the changes $|1'''1'| - |1''1| = (\partial u_1'/\partial r)\,dr$ and $|1'2'| - |12| = u_1' d\varphi$, $|1'3'| - |13| = u_1' d\nu$ are a consequence of both the radial stress σ_{11}' and the tangential stresses σ_{22}', σ_{33}' (see Figure 3.5), respectively. The points 1, $1'$, $1''$, $1'''$ are situated in the planes $x_1'x_2'$ and $x_1'x_3'$ (see Figure 3.1).

3.6. Equilibrium Equations

Solid continuum mechanics considers a condition of equilibrium of forces, which act on sides of an infinitesimal part of a solid continuum. The equilibrium equations of the forces, which act on sides of the infinitesimal spherical cap, are related to the axes x_1', x_2', x_3' at the arbitrary point P (see Figure 3.1).

In case of the axis x_1' (see Figure 3.5), we get

$$\left(\sigma_{11}' + \frac{\partial \sigma_{11}'}{\partial r}\,dr\right)(r+dr)\,d\varphi\,(r+dr)\,d\nu$$
$$+\left(\sigma_{12}' + \frac{\partial \sigma_{12}'}{\partial \varphi}\,d\varphi\right)\cos\frac{d\varphi}{2}\,r\,d\nu\,dr + \left(\sigma_{13}' + \frac{\partial \sigma_{13}'}{\partial \nu}\,d\nu\right)\cos\frac{d\nu}{2}\,r\,d\varphi\,dr$$
$$-\left[\sigma_{11}'\,r\,d\varphi\,r\,d\nu + \left(\sigma_{22}' + \frac{\partial \sigma_{22}'}{\partial \varphi}\,d\varphi\right)\sin\frac{d\varphi}{2}\,r\,d\nu\,dr\right.$$

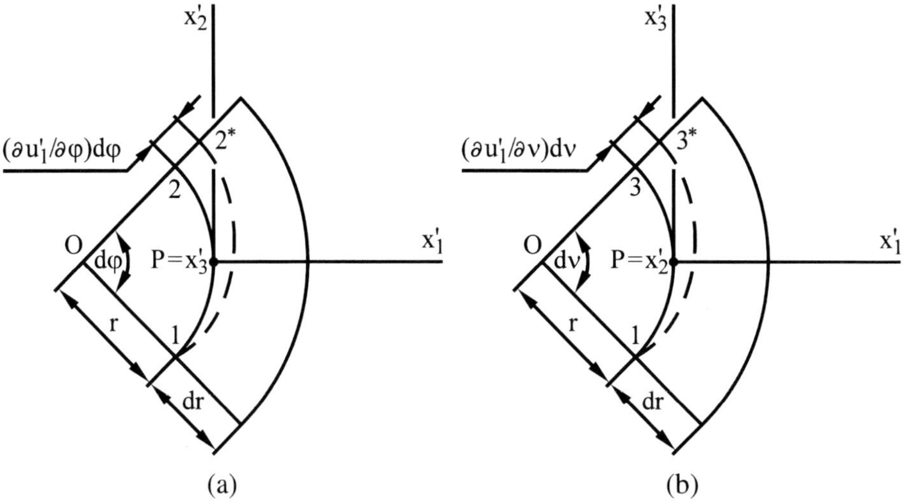

Figure 3.4. The radial displacement $u'_1 = u'_1(\varphi, \nu)$ of the infinitesimal spherical cap at the point P (see Figure 3.1) in the planes (a) $x'_1 x'_2$ and (b) $x'_1 x'_3$ (see Figure 3.1). The changes $|22^*| = (\partial u'_1/\partial\varphi)\,d\varphi$ and $|33^*| = (\partial u'_1/\partial\nu)\,d\nu$ of the function $u'_1 = u'_1(\varphi, \nu)$ of the variables φ, ν are effected by the shear stresses $\sigma'_{12} = \sigma'_{21}$ and $\sigma'_{13} = \sigma'_{31}$ (see Figure 3.5), respectively.

$$+\left(\sigma'_{33} + \frac{\partial\sigma'_{33}}{\partial\nu}d\nu\right)\sin\frac{d\nu}{2}r\,d\varphi\,dr + \sigma'_{22}\sin\frac{d\varphi}{2}r\,d\nu\,dr$$
$$\left. + \sigma'_{33}\sin\frac{d\nu}{2}r\,d\varphi\,dr + \sigma'_{12}\cos\frac{d\varphi}{2}r\,d\nu\,dr + \sigma'_{13}\cos\frac{d\nu}{2}r\,d\varphi\,dr\right] = 0. \quad (3.24)$$

In case of the axis x'_2 (see Figure 3.5), we get

$$\left(\sigma'_{22} + \frac{\partial\sigma'_{22}}{\partial\varphi}d\varphi\right)\cos\frac{d\varphi}{2}r\,d\nu\,dr + \left(\sigma'_{21} + \frac{\partial\sigma'_{21}}{\partial r}dr\right)(r+dr)\,d\varphi\,(r+dr)\,d\nu$$
$$+\left(\sigma'_{12} + \frac{\partial\sigma'_{12}}{\partial\varphi}d\varphi\right)\sin\frac{d\varphi}{2}r\,d\nu\,dr + \sigma'_{12}\sin\frac{d\varphi}{2}r\,d\nu\,dr$$
$$-\left(\sigma'_{22}\cos\frac{d\varphi}{2}r\,d\nu\,dr + \sigma'_{21}r\,d\varphi\,r\,d\nu\right) = 0. \quad (3.25)$$

In case of the axis x'_3 (see Figure 3.5), we get

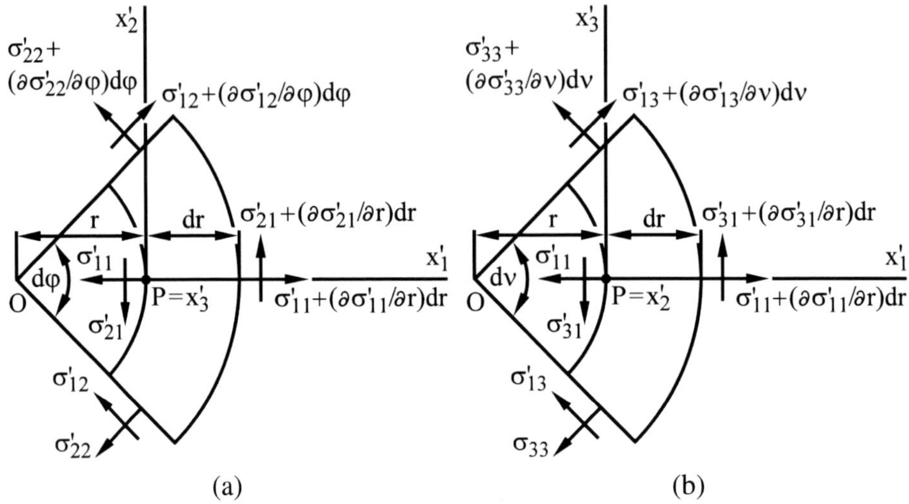

Figure 3.5. Cross-sections of the infinitesimal spherical cap in the planes (a) $x_1'x_2'$ and (b) $x_1'x_3'$. The radial stress σ_{11}', the tangential stresses σ_{22}', σ_{33}', the shear stresses $\sigma_{12}' = \sigma_{21}'$, $\sigma_{13}' = \sigma_{31}'$, along with changes of all these stresses, acting on sides of the infinitesimal spherical cap at the arbitrary point P (see Figure 3.1).

$$\left(\sigma_{33}' + \frac{\partial\sigma_{33}'}{\partial\nu}d\nu\right)\cos\frac{d\nu}{2}\,r\,d\varphi\,dr + \left(\sigma_{31}' + \frac{\partial\sigma_{31}'}{\partial r}dr\right)(r+dr)\,d\varphi\,(r+dr)\,d\nu$$
$$+\left(\sigma_{13}' + \frac{\partial\sigma_{13}'}{\partial\nu}d\nu\right)\sin\frac{d\nu}{2}\,r\,d\varphi\,dr + \sigma_{13}'\sin\frac{d\nu}{2}\,r\,d\varphi\,dr$$
$$-\left(\sigma_{33}'\cos\frac{d\nu}{2}\,r\,d\varphi\,dr + \sigma_{31}'\,r\,d\varphi\,r\,d\nu\right) = 0. \tag{3.26}$$

Due to $d\varphi \approx 0$, $d\nu \approx 0$, $dr \approx 0$, we get $\sin(d\varphi/2) \approx d\varphi/2$, $\sin(d\nu/2) \approx d\nu/2$, $\cos(d\varphi/2) = \cos(d\nu/2) \approx 1$, $(dr)^2 = (d\varphi)^2 = (d\nu)^2 = 0$ [25]. The equilibrium equations (3.24)–(3.26) are transformed to the forms

$$2\sigma_{11}' - \sigma_{22}' - \sigma_{33}' + r\frac{\partial\sigma_{11}'}{\partial r} + \frac{\partial\sigma_{12}'}{\partial\varphi} + \frac{\partial\sigma_{13}'}{\partial\nu} = 0, \tag{3.27}$$

$$\frac{\partial\sigma_{22}'}{\partial\varphi} + 3\sigma_{12}' + r\frac{\partial\sigma_{12}'}{\partial r} = 0, \tag{3.28}$$

$$\frac{\partial \sigma'_{33}}{\partial \nu} + 3\sigma'_{13} + r\frac{\partial \sigma'_{13}}{\partial r} = 0. \tag{3.29}$$

As presented in Section 3.3, the infinitesimal spherical cap exhibits the radial displacement u'_1 along the axis x'_1 (see Figure 3.1). This radial displacement is a consequence of the radial stress p_p, which acts at the particle-matrix boundary. Due to this radial stress, we get $\sigma'_{23} = 0$, where σ'_{23} is a shear stress.

3.7. Hooke's Law

With regard to $\varepsilon'_{23} = 0$, $\sigma'_{23} = 0$ (see Sections 3.5, 3.6), Hooke's law for an isotropic elastic solid continuum is derived as [6]

$$\varepsilon'_{11} = s_{11}\sigma'_{11} + s_{12}\left(\sigma'_{22} + \sigma'_{33}\right), \tag{3.30}$$

$$\varepsilon'_{22} = s_{12}\left(\sigma'_{11} + \sigma'_{33}\right) + s_{11}\sigma'_{22}, \tag{3.31}$$

$$\varepsilon'_{33} = s_{12}\left(\sigma'_{11} + \sigma'_{22}\right) + s_{11}\sigma'_{33}, \tag{3.32}$$

$$\varepsilon'_{13} = s_{44}\sigma'_{13}, \tag{3.33}$$

$$\varepsilon'_{12} = s_{44}\sigma'_{12}. \tag{3.34}$$

The elastic moduli s_{11}, s_{12}, s_{44} have the forms [6]

$$s_{11} = \frac{1}{E}, \quad s_{12} = -\frac{\mu}{E}, \quad s_{44} = \frac{2(1+\mu)}{E}, \tag{3.35}$$

where E and μ represent Young's modulus and Poisson's ratio, respectively.

In case of an isotropic elastic solid continuum, we get $\mu = 0.25$ [6]. In case of a real isotropic material, we get $\mu < 0.5$ [18]. Young' modulus and Poisson's ratio are related to the spherical particle ($q = p$) and matrix ($q = m$). Consequently, the transformations $E \rightarrow E_q$, $\mu \rightarrow \mu_q$ ($q = p,m$) are required to be considered.

3.8. System of Differential Equations

Let Cauchy's equations (see Equations (3.19)–(3.23) be substituted to Hooke's law (see Equations (3.27)–(3.29)). Consequently, the radial stress σ'_{11}, the tangential stresses σ'_{22}, σ'_{33} and the shear stresses σ'_{12}, σ'_{13} have the forms

$$\sigma'_{11} = (c_1 + c_2)\frac{\partial u'_1}{\partial r} - 2c_2\frac{u'_1}{r}, \tag{3.36}$$

$$\sigma'_{22} = \sigma'_{33} = -c_2\frac{\partial u'_1}{\partial r} + c_1\frac{u'_1}{r}, \tag{3.37}$$

$$\sigma'_{12} = \frac{1}{s_{44}r}\frac{\partial u'_1}{\partial \varphi}, \tag{3.38}$$

$$\sigma'_{13} = \frac{1}{s_{44}r}\frac{\partial u'_1}{\partial \nu}, \tag{3.39}$$

where the condition $\sigma'_{22} = \sigma'_{33}$ results from $\varepsilon'_{22} = \varepsilon'_{33}$ (see Equations (3.20), (3.21), (3.31), (3.32)). The coefficient c_i $(i = 1,2,3)$ is derived as (see Equation (3.35))

$$c_1 = \frac{s_{11}}{s_{11}(s_{11} + s_{12}) - 2s_{12}^2} = \frac{E}{(1+\mu)(1-2\mu)},$$
$$c_2 = \frac{s_{12}c_1}{s_{11}} = -\frac{\mu E}{(1+\mu)(1-2\mu)}, \quad c_3 = \frac{s_{44}c_1 + 2}{s_{44}c_2 - 1} = -4(1-\mu) \tag{3.40}$$

In case of an isotropic elastic solid continuum, we get $\mu = 0.25$ [6]. In case of a real isotropic material, we get $\mu < 0.5$ [18]. Due to $\mu < 0.5$, we get $c_3 < 0$. The coefficient c_3, which is an exponent of the function $u_r = r^{c_3}$, is considered within the mathematical procedures in Sections 7.1, 8.1 (see Equations (7.4), (7.6), (8.4), (8.6)).

Substituting Equations (3.44)–(3.48) to Equation (3.27) and to the sum $[\partial \text{Eq.}(3.28)/\partial\varphi] + [\partial \text{Eq.}(3.29)/\partial\nu]$, the equilibrium equations (3.27)–(3.29) are transformed to the forms

$$r^2\frac{\partial^2 u'_1}{\partial r^2} + 2r\frac{\partial u'_1}{\partial r} - 2u'_1 + \frac{U'_1}{s_{44}(c_1 + c_2)} = 0, \tag{3.41}$$

$$r\,\frac{\partial U_1'}{\partial r} = c_3\, U_1', \tag{3.42}$$

where the function U_1' is derived as

$$U_1' = \frac{\partial^2 u_1'}{\partial \varphi^2} + \frac{\partial^2 u_1'}{\partial \nu^2}. \tag{3.43}$$

Solutions of the differential equations (3.36), (3.37) are determined by the different mathematical procedures in Sections 6.1, 7.1, 8.1.

The solution for $u_1' = u_1'(r, \varphi, \nu)$ (see Equation (6.7)–(6.9)), which results from the mathematical procedures in Section 6.1, includes the integration constants C_1, C_2. These constants are determined by the boundary conditions (5.1)–(5.6).

The solutions for $u_1' = u_1'(r, \varphi, \nu)$ (see Equations (7.4)–(7.7), (8.4)–(8.7)), which result from the mathematical procedures in Sections 7.1, 8.1, include the integration constants C_1, C_2, C_3. These constants are determined by the boundary conditions (5.1)–(5.8), where the analysis in Section 7.1 is considered (see Items 1, 2, 3, 4, page 53).

With regard to Section 3.9, such a solution of these different mathematical procedures is considered to exhibit a minimum value of the elastic energy W_c (see Equations (3.44)–(3.49)). This minimum value is determined numerically for a real two-component material with isotropic components. This numerical determination considers Equations (9.38), (9.39).

3.9. Elastic Energy

The elastic energy density w_q $(q = p, m, v)$, which is accumulated at the arbitrary point in the spherical particle and the matrix of the cubic cell *I*, *II*, has the form [6]

$$w_q = \frac{1}{2}\left(\sum_{i=1}^{3} \sigma_{iiq}' \varepsilon_{iiq}' + \sum_{i,j=1;\ i\neq j}^{3} \sigma_{ijq}' \varepsilon_{ijq}'\right), \quad q = p, m, v. \tag{3.44}$$

With regard to Equations (3.19)–(3.23), (3.36)–(3.39), we get

$$w_q = \frac{1}{2}(c_{1q} + c_{2q})\left(\frac{\partial u'_{1q}}{\partial r}\right)^2 - \frac{2c_{1q}u'_{1q}}{r}\frac{\partial u'_{1q}}{\partial r} + \frac{c_{1q}\left(u'_{1q}\right)^2}{r^2}$$
$$+ \frac{1}{s_{44q}r^2}\left[\left(\frac{\partial u'_{1q}}{\partial \varphi}\right)^2 + \left(\frac{\partial u'_{1q}}{\partial \nu}\right)^2\right], \quad q = p, m, v. \tag{3.45}$$

where c_{iq} $(i = 1,2,3)$ is given by Equation (3.40). The subscript $q = p$ is related to the spherical particle. The subscripts $q = m$ and $q = v$ are related to the matrix of the cubic cell *I* and *II*, respectively.

As presented in Section 3.2, the intervals $\varphi \in \langle 0, \pi/2\rangle$ and $\nu \in \langle 0, \pi/2\rangle$ are sufficient to be considered. In case of the spherical coordinates (r, φ, ν) (see Figure 3.1), the elastic energy W_p, which is accumulated in the volume V_p of the spherical particle, is derived as

$$W_p = \int_{V_p} w_p\, dV = 8\int_0^{\pi/2}\int_0^{\pi/2}\int_0^{R_p} w_p r^2\, dr\, d\varphi\, d\nu, \tag{3.46}$$

where $dV = r^2 dr\, d\varphi\, d\nu$ is volume of the infinitesimal spherical cap (see Figure 3.1).

The elastic energy W_m, which is accumulated in the volume V_m of the matrix of the cubic cell *I* (see Figure 2.1), has the form

$$W_m = \int_{V_m} w_m\, dV = 8\int_0^{\pi/2}\int_0^{\pi/2}\int_{R_p}^{r_c} w_m r^2\, dr\, d\varphi\, d\nu. \tag{3.47}$$

The elastic energy W_v, which is accumulated in the volume V_v of the matrix of the cubic cell *II* (see Figure 2.1), is derived as

$$W_v = \int_{V_v} w_v\, dV = 8\int_0^{\pi/2}\int_0^{\pi/2}\int_{R_v}^{r_c} w_v r^2\, dr\, d\varphi\, d\nu. \tag{3.48}$$

The multi-particle-matrix system consists of the cubic cells *1234* (see Figure 2.1b). The definition cell *1234* consists of one cubic cells of the type *I* and one cubic cell of the type *II*, i.e., *1234* = *I* +*II*. The elastic energy W_c, which is accumulated in the definition cell *1234*, has the form

$$W_c = W_p + W_m + W_v. \tag{3.49}$$

As presented in Section 2.1, the multi-particle-matrix is defined by the cell *1234* (see Figure 2.1b). Accordingly, W_c represents elastic energy of the multi-particle-matrix system.

As presented in Sections 6.1, 7.1, 8.1, the thermal and phase-transformation-induced stresses are determined by different mathematical procedures. These different procedures result in different partial differential equations with non-zero right-hand sides (see Equations (6.6), (7.3), (8.3)). These differential equations result in different mathematical solutions for the thermal and phase-transformation-induced stresses (see Equations (6.7)–(6.9), (7.4)–(7.10), (8.4)–(8.10)). These different solutions are determined for the spherical particle and the cell matrix *I*, *II* (see Figure 2.1), where the mathematical boundary conditions in Chapter 5 are considered. Due to these different solutions, the principle of minimum total potential energy is required to be considered [6]. This principle means that an elastic solid body (i.e., the multi-particle-matrix system) tends to exhibit a minimum value of the total potential energy W_t [6].

Let W_d represent deformation energy. Let W_v represent energy, which is induced by the volume force $\overrightarrow{F_v}$. Let W_s represent energy, which is induced by the surface force $\overrightarrow{F_s}$. As presented in [6], we get $W_t = W_d - (W_v + W_s)$. In case of the multi-particle-matrix system (see Figure 2.1), we get $\overrightarrow{F_v} = \overrightarrow{F_s} = 0$, and then $W_v = W_s = 0$, $W_t = W_d$. Additionally, in case of the multi-particle-matrix system, we get $W_d = W_c$, and then $W_t = W_d = W_c$ (see Equation (3.44)–(3.49)). Due to $W_t = W_c$, the principle of minimum total potential energy is 'transformed' to the principle of a minimum value of the elastic energy W_c. Accordingly, such solutions for for the spherical particle and the cell matrix *I*, *II* are considered to exhibit a minimum value of W_c (see Equation (3.44)–(3.49)), which is accumulated in the definition cell *1234* (see Figure 2.1b).

Chapter 4

Reason of Stresses

The thermal stresses are a consequence of the condition $\alpha_p \neq \alpha_m$, where α_p and α_p are thermal expansion coefficients of the spherical particle and matrix of the multi-particle-matrix system (see Figure 2.1).

Let a phase transformation originate at the temperature T_{tq}, where $T_{tq} \in \langle T_f, T_r \rangle$; T_f is final temperature of a cooling process; and T_r is relaxation temperature of the multi-particle-matrix system. Let the phase transformation originate at least in one component of the multi-particle-matrix system. This transformation is a reason of the phase-transformation-induced strain ε_{tq}, where the subscripts $q = p$ and $q = m$ are related to the spherical particle and matrix, respectively.

4.1. Relaxation Temperature

If $T \geq T_r$, then the thermal and phase-transformation-induced stresses are relaxed by thermal-activated processes [18]. The relaxation temperature T_r is defined by the formula $T_r = (0.35 - 0.4) \times T_m$ [18], where T_m is melting temperature of a material.

Let a two-component material consist of precipitates, which are distributed in crystal grains (see Item 1, Section2.1). If the precipitates are formed in a liquid matrix, then T_m represents a minimum value of the set $\{T_{mp}, T_{mm}\}$, where T_{mp} and T_{mm} is melting temperature of the precipitates and matrix, respectively. If the precipitates are formed in a solid matrix with the melting temperature T_{mm}, then we get $T_m = T_{mm}$.

If a two-component material consist of two types of crystal grains (see Item 2, Section 2.1), then T_m represents melting temperature of this material.

4.2. Coefficient β_q

If the phase transformation originates at the temperature $T_{tq} \in \langle T_f, T_r \rangle$, then the coefficient $\beta_q = \beta_q(T)$ at the temperature $T \in \langle T_f, T_{tq} \rangle \subset \langle T_f, T_r \rangle$ has the form

$$\beta_q = \varepsilon_{tq} + \int_{T_{tq}}^{T_r} \alpha_q^{(1)} \, dT + \int_{T}^{T_{tq}} \alpha_q^{(2)} \, dT,$$
$$T_{tq} \in \langle T_f, T_r \rangle, \quad T \in \langle T_f, T_{tq} \rangle \subset \langle T_f, T_r \rangle, \tag{4.1}$$

where $\alpha_q^{(1)} = \alpha_q^{(1)}(T)$ and $\alpha_q^{(2)} = \alpha_q^{(2)}(T)$ represent a thermal expansion coefficient of the component at the temperature $T \geq T_{tq}$ and $T \leq T_{tq}$, respectively. The subscripts $q = p$ and $q = m$ are related to the *isotropic* spherical particle and *isotropic* matrix of the multi-particle-matrix system (see Figure 2.1). If $\alpha_q^{(2)}$ and $\alpha_q^{(1)}$ are not functions of $T \leq T_{tq}$ and $T \geq T_{tq}$ (i.e., $\alpha_q^{(2)} \neq f(T)$, $\alpha_q^{(1)} \neq f(T)$), respectively, then we get

$$\beta_q = \varepsilon_{tq} + \alpha_q^{(1)}(T_r - T_{tq}) + \alpha_q^{(2)}(T_{tq} - T),$$
$$T_{tq} \in \langle T_f, T_r \rangle, \quad T \in \langle T_f, T_{tq} \rangle \subset \langle T_f, T_r \rangle. \tag{4.2}$$

If $T > T_{tq}$, then the coefficient $\beta_q = \beta_q(T)$ at the temperature $T \in (T_{tq}, T_r\rangle \subset \langle T_f, T_r \rangle$ has the form

$$\beta_q = \int_{T}^{T_r} \alpha_q^{(1)} \, dT, \quad T_{tq} \in \langle T_f, T_r \rangle, \quad T \in (T_{tq}, T_r\rangle \subset \langle T_f, T_r \rangle. \tag{4.3}$$

If $\alpha_q^{(1)} \neq f(T)$ for $T \geq T_{tq}$, then we get

$$\beta_q = \alpha_q^{(1)}(T_r - T), \quad T \in (T_{tq}, T_r\rangle \subset \langle T_f, T_r \rangle. \tag{4.4}$$

If the phase transformation does not originate at the temperature $T_{tq} \in \langle T_f, T_r \rangle$ (i.e., $T_{tq} \notin \langle T_f, T_r \rangle$), then the coefficient $\beta_q = \beta_q(T)$ at the temperature $T \in \langle T_f, T_r \rangle$ has the form

$$\beta_q = \int_T^{T_r} \alpha_q dT, \quad T_{tq} \notin \langle T_f, T_r \rangle, \quad T \in \langle T_f, T_r \rangle, \tag{4.5}$$

where $\alpha_q = \alpha_q(T)$ is a thermal expansion coefficient on the condition $T_{tq} \notin \langle T_f, T_r \rangle$. If $\alpha_q \neq f(T)$ for $T \in \langle T_f, T_r \rangle$, then we get

$$\beta_q = \alpha_q (T_r - T), \quad T_{tq} \notin \langle T_f, T_r \rangle, \quad T \in \langle T_f, T_r \rangle. \tag{4.6}$$

4.3. Phase-Transformation-Induced Strain ε_{tq}

As presented in Preface, the thermal and phase-transformation-induced stresses are determined for *isotropic* components of the multi-particle-matrix system in Figure 2.1. The isotropic components are characterized by cubic crystal lattices. The cubic crystal lattices (CCL) exhibit [18]

- a simple modification ($K6$), which is characterized by atoms at corner points of CCL,
- a body-centered modification ($K8$), which is characterized by an atom at an intersection point of diagonals of CCL (i.e., at a center of CCL), as well as by atoms at corner points of CCL,
- a face-centered modification ($K12$), which is characterized by a central atom on each side of CCL, as well as by atoms at corner points of CCL.

Let a phase transformation originate during a cooling process. Let T_{tq} represent temperature of this phase transformation, where $T_{tq} \in \langle T_f, T_r \rangle$. The phase transformation results in the transformation $a_q^{(1)} \rightarrow a_q^{(2)}$, where $a_q^{(1)}$ and $a_q^{(2)}$ represent dimensions of cubic crystal lattices at the temperature $T \geq T_{tq}$ and $T \leq T_{tq}$, respectively. Finally, the transformation $a_q^{(1)} \rightarrow a_q^{(2)}$ results in the phase-transformation-induced strain ε_{tq}, which is derived as

$$\varepsilon_{tq} = \frac{a_q^{(2)} - a_q^{(1)}}{a_q^{(1)}}, \tag{4.7}$$

where the subscripts $q = p$ and $q = m$ are related to the spherical particle and matrix of the multi-particle-matrix system in Figure 2.1, respectively. This strain is a reason of the phase-transformation-induced stresses.

Chapter 5

Mathematical Boundary Conditions

The integration constants C_1, C_2, C_3, which are included in Equations (6.7), (7.4), (8.4), are determined by the mandatory and additional boundary conditions (5.1)–(5.6) and (5.7)–(5.10), respectively.

5.1. Spherical Particle

The absolute values $\left|u'_{1p}\right|$, $\left|\varepsilon'_{ijp}\right|$, $\left|\sigma'_{ijp}\right|$ $(i, j = 1,2,3)$ are required to represent increasing functions of $r \in \langle 0, R_p \rangle$ (see Equations (3.19)–(3.23), (3.36)–(3.39)). These increasing functions exhibit maximum values on the particle-matrix boundary, i.e., for $r = R_p$. Additionally, the conditions $\left(u'_{1p}\right)_{r \to 0} \nrightarrow \pm\infty$, $\left(\varepsilon_{ijp}\right)_{r \longrightarrow 0} \nrightarrow \pm\infty$, $\left(\sigma_{ijp}\right)_{r \to 0} \nrightarrow \pm\infty$ are required to be valid.

The boundary conditions for the radial stress σ'_{11} and the radial displacement u'_1 have the forms

$$\left(\sigma'_{11p}\right)_{r=R_p} = -p_p, \tag{5.1}$$

$$\left(u'_{1p}\right)_{r=0} = 0. \tag{5.2}$$

Equations (5.1) and (5.2) represent stress and geometric boundary conditions, respectively. The radial stress p_p, which acts at the particle-matrix boundary is determined by the formula (5.12). The distance r_c is determined in Section 3.2 (see Equations (3.1)–(3.3), (3.8), (3.15)–(3.18)).

5.2. Cell Matrix Around Particle

The absolute values $|u'_{1m}|$, $\left|\varepsilon'_{ijm}\right|$, $\left|\sigma'_{ijm}\right|$ $(i, j = 1,2,3)$ are required to represent decreasing functions of $r \in \langle R_p, r_c\rangle$ (see Equations (3.19)–(3.23), (3.36)–(3.39)). These decreasing functions exhibit maximum values on the particle-matrix boundary, i.e., for $r = R_p$. These decreasing courses are ensured by the integration constants $C_{1m} \neq 0$, $C_{2m} \neq 0$, which are determined by mandatory boundary conditions. These conditions are derived as

$$\left(\sigma'_{11m}\right)_{r=R_p} = -p_p, \tag{5.3}$$

$$\left(u'_{1m}\right)_{r=r_c} = u'_{1cm} = u_{1c0}\, f_c, \tag{5.4}$$

where the function $f_c = f_c(\varphi, \nu)$ is given by Equations (3.16, (3.18), and u_{1c0} is a radial displacement along the axis x_i $(i = 1,2,3)$ on the cell surface *I,II*, i.e., for $r = d/2$ (see Figure 2.1). The displacement u_{1c0} is determined by the analysis in Section 4.6. Equations (5.3) and (5.4) represent stress and geometric boundary conditions, respectively.

5.3. Cell Matrix Around Void

The absolute values $|u'_{1v}|$, $\left|\varepsilon'_{ijv}\right|$, $\left|\sigma'_{ijv}\right|$ $(i, j = 1,2,3)$ are required to represent increasing functions of $r \in \langle R_v, r_c\rangle$ (see Equations (3.19)–(3.23), (3.36)–(3.39)). These increasing functions exhibit maximum values on the cell surface, i.e., for $r = r_c$ (see Equations (3.15)–(3.18)). These increasing courses are ensured by the integration constants $C_{1v} \neq 0$, $C_{2v} \neq 0$, which are determined by mandatory boundary conditions. These conditions have the forms

$$\left(\sigma'_{11v}\right)_{r=R_v} = 0, \tag{5.5}$$

$$\left(u'_{1v}\right)_{r=r_c} = -u'_{1cv} = -u_{1c0}\, f_c, \tag{5.6}$$

where the displacement u_{1c0} along the axis x_i $(i = 1,2,3)$ on the cell surface *I,II* (i.e., for $r = d/2$) is determined by the analysis in Section 4.6 (see Figure 2.1).

5.4. Additional Boundary Condition

As presented in Section 2.1, the multi-particle-matrix system is imaginary divided into cubic cells (see Figure 2.1). The surface *1234* of the neighbouring cubic cells A and B in Figure 5.1 is not a physical boundary.[1] The functions $u'_{1A} = u'_{1A}(r, \varphi_A, \nu_A)$ and $u'_{1B} = u'_{1B}(r, \varphi_B, \nu_B)$ are related to the cell A and B, respectively. These functions are connected at the point C. This connection is assumed to be 'smooth'. Due to this assumption, the functions u'_{1A} and u'_{1B} of the variable r are required not to mutually create a singular connection at the point C. Accordingly, C is required not to represent a singular point.

With regard to the cubic cells *I, II*, we get $u'_{1A} = u'_{1m}$ and $u'_{1B} = u'_{1v}$. To fulfil this non-singularity assumption, the functions $u'_{1m} = u'_{1m}(r, \varphi, \nu)$ and $u'_{1v} = u'_{1v}(r, \varphi, \nu)$ are required to be extremal on the cell surface, i.e., for $r = r_c + u'_{1c}$ and $r = r_c - u'_{1c}$, respectively. This extreme for for $r = r_c + u'_{1c}$ and $r = r_c - u'_{1c}$ represents a minimum and maximum due to a decreasing and increasing course of the functions $|u'_{1m}|$ and $|u'_{1v}|$ of the variable $r \in \langle R_p, r_c \rangle$ and $r \in \langle R_v, r_c \rangle$, respectively. With regard to Equation (3.19), the additional boundary conditions for the cell matrix around the spherical particle and void are derived as

$$\left(\varepsilon'_{11m}\right)_{r=r_c+u'_{1c}} = \left(\frac{\partial u'_{1m}}{\partial r}\right)_{r=r_c+u'_{1c}} = 0, \tag{5.7}$$

$$\left(\varepsilon'_{11v}\right)_{r=r_c-u'_{1c}} = \left(\frac{\partial u'_{1v}}{\partial r}\right)_{r=r_c-u'_{1c}} = 0. \tag{5.8}$$

On the one hand, this non-singularity assumption can be valid for the elastic energy density $w_q = w_q(r, \varphi, \nu)$ $(q = p, v)$ (see Equation (3.44)). Accordingly, we get

$$\left(\frac{\partial w_m}{\partial r}\right)_{r=r_c+u'_{1c}} = 0, \tag{5.9}$$

$$\left(\frac{\partial w_v}{\partial r}\right)_{r=r_c-u'_{1c}} = 0. \tag{5.10}$$

On the other hand, the following analysis is required to be considered. With regard to Equation (3.44), the elastic energy density $w_q = w_q(r, \varphi, \nu) > 0$

[1] In contrast to *1234*, a surface of two neighbouring material components represents a physical boundary, e.g., the particle-matrix boundary in Figure 2.1.

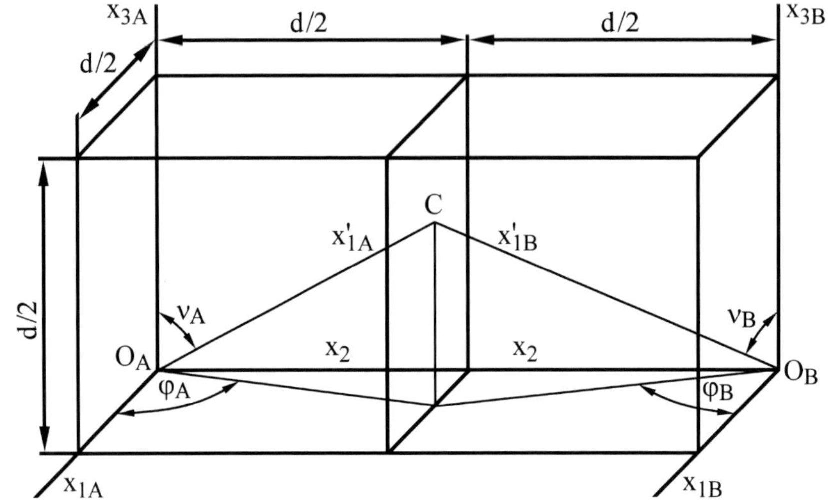

Figure 5.1. The arbitrary point C on the surface *1234* of the neighbouring cubic cells A and B with the centres O_A and O_B (see Figure 2.1a), respectively, where $\varphi_A = \varphi_B = \varphi$, $\nu_A = \nu_B = \nu$, $r_{cA} = \left|\overline{O_A C}\right| = r_{cB} = \left|\overline{O_B C}\right| = r_c$. The distance r_c is determined in Section 3.2 (see Equations (3.1)–(3.3), (3.8), (3.15)–(3.18)).

$(q = p, v)$ includes the shear stresses σ'_{12q} and σ'_{13q}, as well as the shear strains ε'_{12q} and ε'_{13q}, which are functions of $\partial u'_{1q}/\partial\varphi$ and $\partial u'_{1q}/\partial\nu$, respectively (see Equations (3.22), (3.23), (3.33), (3.34)). With regard to Equations (5.9), (5.10), (6.7), (7.4), (8.4), the derivatives $\partial u'_{1q}/\partial\varphi$ and $\partial u'_{1q}/\partial\nu$ are functions of $\partial C_{iq}/\partial\varphi$ and $\partial C_{iq}/\partial\nu$ $(i = 1, \ldots, \mathrm{n})$, respectively, where n is a number of the boundary conditions. Accordingly, the elastic energy density $w_q = w_q(r, \varphi, \nu)$ and Equations (5.9), (5.10) include the derivatives $\partial C_{iq}/\partial\varphi$, $\partial C_{iq}/\partial\nu$ $(i = 1, \ldots, \mathrm{n})$.

In case of the cell matrix around the spherical particle, the boundary conditions (5.3), (5.4), (5.9) are considered (i.e., $n = 3$). Consequently, the integration constants C_{1m}, C_{2m}, C_{3m}, along with the derivatives $\partial C_{1m}/\partial\varphi$, $\partial C_{1m}/\partial\nu$, $\partial C_{2m}/\partial\varphi$, $\partial C_{2m}/\partial\nu$, $\partial C_{3m}/\partial\varphi$, $\partial C_{3m}/\partial\nu$, are required to be determined by nine boundary conditions for the cell matrix around the spherical particle. Consequently, six additional boundary conditions are required to be determined, but no additional boundary conditions exist.

If the boundary conditions (5.3), (5.4), (5.7), (5.9) are considered (i.e., $n = 4$), then the integration constants C_{1m}, C_{2m}, C_{3m}, C_{4m}, along with the derivatives $\partial C_{1m}/\partial\varphi$, $\partial C_{1m}/\partial\nu$, $\partial C_{2m}/\partial\varphi$, $\partial C_{2m}/\partial\nu$, $\partial C_{3m}/\partial\varphi$, $\partial C_{3m}/\partial\nu$, $\partial C_{4m}/\partial\varphi$, $\partial C_{4m}/\partial\nu$, are required to be determined by twelve boundary conditions. Consequently, eight additional boundary conditions are required to be determined, but no additional boundary conditions exist.

The same analysis is valid for the cell matrix around the spherical void. Finally, the additional boundary conditions (5.9), (5.10) can not be considered.

5.5. Determination of Radial Stress p_p

The thermal and phase-transformation-induced stresses in the multi-particle-matrix system (see Figure 2.1) are a consequence of the radial stress p_p, which acts at the particle-matrix boundary (i.e., for $r = R_p$). If $\beta_m > \beta_p$, then p_p is compressive, where $p_p > 0$. If $\beta_m < \beta_p$, then p_p is tensile, where $p_p < 0$. The coefficient β_q ($q = p,m$) is determined in Section 4.2.

The radial displacement $\left(u'_{1q}\right)_{r=R_p}$ in the spherical particle and the cell matrix I at the particle-matrix boundary is derived as

$$\left(u'_{1q}\right)_{r=R_p} = \beta_q R_p, \quad q = p,m. \tag{5.11}$$

With regard to Equations (3.20), (3.21), the radial stress p_p is determined by the following formula

$$\left(\varepsilon'_{22m}\right)_{r=R_p} - \left(\varepsilon'_{22p}\right)_{r=R_p} = \beta_m - \beta_p. \tag{5.12}$$

5.6. Determination of Displacement u_{1c0}

The displacement u_{1c0} along the axis x_i ($i = 1,2,3$) on the cell surface *I, II* (i.e., for $r = d/2$) is determined by the following energy analysis (see Figure 2.1).

As presented in Section 2.1, the multi-particle-matrix system is defined by the definiton cell *1234* (see Figure 2.1). This cell consists of one cubic cell *I* and one cubic cell *II*. Total potential energy of the multi-particle-matrix system is represented by the elastic energy W_c (see Section 3.8), which is accumulated in this definiton cell, where $W_c = W_p + W_m + W_v$ (see Equations (3.44)–(3.49)). The elastic energy $W_I = W_p + W_m$ is accumulated in the cubic cell *I*. The elastic energy $W_{II} = W_v$ is accumulated in the cubic cell *II*.

With regard to $(u'_{1m})_{r=r_c} = -(u'_{1v})_{r=r_c}$ (see Equations (5.4), (5.6)), the difference $W_I - W_{II}$ is required to be considered. Consequently, the thermal and phase-transformation-induced stresses, which act in this definiton cell, are required to fulfilled the following energy condition

$$\frac{\partial (W_p + W_m - W_v)}{\partial u_{1c0}} = 0. \tag{5.13}$$

Finally, the displacement u_{1c0} is determined by the condition (5.13).

Chapter 6

Analytical Model 1

The mathematical procedure in Section 6.1, which is applied to the differential equations (3.41), (3.42), results in the differential equation (6.6) with a non-zero right-hand side. Consequently, Wronskian's method is applied to Equation (6.6) [25]. The solution of Equation (6.6) includes the integration constants C_1, C_2 (see Equation (6.7)), which are determined by the boundary conditions (5.1)–(5.6).

6.1. Mathematical Procedure 1

Performing $r\left[\partial \mathrm{Eq.}(3.42)/\partial r\right]$, the differential equation (3.42) is transformed to the form

$$r^2 \frac{\partial^2 U_1'}{\partial r^2} + (1 - c_3)\, r \frac{\partial U_1'}{\partial r} = 0, \tag{6.1}$$

where the coefficient $c_3 < 0$ and the function $U_1' = U_1'(r, \varphi, \nu)$ are given by Equations (3.40) and (3.43), respectively. Substituting Equation (3.42) to Equation (6.1), the differential equation (6.1) is derived as

$$r^2 \frac{\partial^2 U_1'}{\partial r^2} + c_3 (1 - c_3)\, U_1' = 0. \tag{6.2}$$

If U_1' is assumed in the form $U_1' = r^{\lambda}$ (see Equation (3.43)), then the solution U_1' of Equation (6.2) has the form

$$U_1' = \frac{\partial^2 u_1'}{\partial \varphi^2} + \frac{\partial^2 u_1'}{\partial \nu^2} = \sum_{i=1}^{2} C_i r^{\lambda_i}. \tag{6.3}$$

With regard to Equation (3.40), the exponent $\lambda_i (i = 1,2)$ and the discriminant D are derived as

$$\lambda_i = \frac{1}{2}\left[1 + (\delta_{1i} - \delta_{2i})\sqrt{D}\right], \quad i = 1,2,$$
$$D = 1 - 4c_3(1 - c_3) = 1 + 16(1 - \mu)[1 + 4(1 - \mu)], \tag{6.4}$$

In case of a real isotropic material, we get $\mu < 0.5$ and $D > 0$ [18]. The coefficient c_3 (see Equation (3.40)) and the real exponents λ_1, λ_2 have the forms

$$c_3 < 0, \quad \lambda_1 > 3, \quad \lambda_2 < -2. \tag{6.5}$$

Substituting Equation (6.3) to Equation (3.41), the differential equation (3.41) is transformed to the form

$$\frac{\partial^2 u_1'}{\partial r^2} + \frac{2}{r}\frac{\partial u_1'}{\partial r} - \frac{2u_1'}{r^2} = -\frac{1}{s_{44}(c_1 + c_2)}\sum_{i=1}^{2} C_i r^{\lambda_i - 2}, \tag{6.6}$$

where the elastic modulus s_{44} and the coefficients c_1, c_2 are given by Equations (3.35) and (3.40), respectively. The integration constant C_i $(i = 1,2,3)$ is determined by the boundary conditions, which are defined in Chapter 5.

Equation (6.6) with the variable r represents a linear differential equation of the second order with respect to the radial displacement u_1'. Due to the non-zero right-hand side, Wronskian's method is applied to Equation (6.6) [25]. Consequently, the radial displacement $u_{1q}' = u_{1q}'(r, \varphi, \nu)$ $(q = p,m,v)$ has the form

$$u_{1q}' = \sum_{i=1}^{2} C_{iq}\,\xi_{iq}\,u_{irq}, \quad q = p,m,v. \tag{6.7}$$

The subscript $q = p$ is related to the spherical particle, where $r \in \langle 0, R_p \rangle$ (see Figure 2.1a). The subscripts $q = m$ and $q = v$ are related to the cell matrix *I* and *II* (see Figure 2.1a), where $r \in \langle R_p, r_c \rangle$ and $r \in \langle R_v, r_c \rangle$, respectively. The function $u_{irq} = u_{irq}(r)$ $(i = 1,2;\ q = p,m,v)$ is derived as

$$u_{irq} = r^{\lambda_{iq}}, \quad i = 1,2; \quad \lambda_{1q} > 3, \quad \lambda_{2q} < -2, \tag{6.8}$$

where $u_{1rq} = u_{1rq}(r)$ and $u_{2rq} = u_{2rq}(r)$ are increasing and decreasing functions of the variable r due to $\lambda_{1q} > 3$ and $\lambda_{2q} < -2$ (see Equation (6.4)), respectively. With regard to Equation (3.40), the coefficient ξ_{iq} ($i = 1,2$; $q = p,m,v$) has the form

$$\xi_{iq} = \frac{1}{3s_{44q}(c_{1q}+c_{2q})}\left(\frac{1}{\lambda_{iq}+2} - \frac{1}{\lambda_{iq}-1}\right)$$
$$= \frac{1-2\mu_q}{6(1-\mu_q)}\left(\frac{1}{\lambda_{iq}+2} - \frac{1}{\lambda_{iq}-1}\right), \quad i = 1,2. \tag{6.9}$$

With regard to Equations (3.19)–(3.23), (3.36)–(3.39), (3.45), (6.7), (6.8), we get

$$\varepsilon'_{11q} = \sum_{i=1}^{2} C_{iq}\xi_{iq}\lambda_{iq} r^{\lambda_{iq}-1}, \tag{6.10}$$

$$\varepsilon'_{22q} = \varepsilon'_{33q} = \sum_{i=1}^{2} C_{iq}\xi_{iq} r^{\lambda_{iq}-1}, \tag{6.11}$$

$$\varepsilon'_{12q} = \sum_{i=1}^{2} \frac{\partial C_{iq}}{\partial \varphi}\xi_{iq} r^{\lambda_{iq}-1}, \tag{6.12}$$

$$\varepsilon'_{13q} = \sum_{i-1}^{2} \frac{\partial C_{iq}}{\partial \nu}\xi_{iq} r^{\lambda_{iq}-1}, \tag{6.13}$$

$$\sigma'_{11q} = \sum_{i=1}^{2} C_{iq}\xi_{iq}\xi_{2+iq} r^{\lambda_{iq}-1}, \tag{6.14}$$

$$\sigma'_{22q} = \sigma'_{33q} = \sum_{i=1}^{2} C_{iq}\xi_{iq}\xi_{4+iq} r^{\lambda_{iq}-1}, \tag{6.15}$$

$$\sigma'_{12q} = \frac{1}{s_{44q}}\sum_{i=1}^{2} \frac{\partial C_{iq}}{\partial \varphi}\xi_{iq} r^{\lambda_{iq}-1}, \tag{6.16}$$

$$\sigma'_{13q} = \frac{1}{s_{44q}}\sum_{i=1}^{2} \frac{\partial C_{iq}}{\partial \nu}\xi_{iq} r^{\lambda_{iq}-1}, \tag{6.17}$$

$$w_q = \sum_{i,j=1}^{2} \xi_{iq}\xi_{jq}\left[C_{iq}C_{jq}\xi_{6+i+2jq} + \frac{1}{s_{44q}}\left(\frac{\partial C_{iq}}{\partial \varphi}\frac{\partial C_{jq}}{\partial \varphi} + \frac{\partial C_{iq}}{\partial \nu}\frac{\partial C_{jq}}{\partial \nu}\right)\right] r^{\lambda_{iq}+\lambda_{jq}-2}, \tag{6.18}$$

where the coefficients ξ_{2+iq}, ξ_{4+iq}, $\xi_{6+i+2jq}$ ($i = 1,2$; $q = p,m,v$) are derived as

$$\xi_{2+iq} = \lambda_{iq}(c_{1q} + c_{2q}) - 2c_{2q}, \quad i = 1,2, \tag{6.19}$$

$$\xi_{4+iq} = c_{1q} - \lambda_{iq}c_{2q}, \quad i = 1,2, \tag{6.20}$$

$$\xi_{6+i+2jq} = \lambda_{iq}\left[\frac{\lambda_{jq}}{2}(c_{1q} + c_{2q}) - 2c_{2q}\right] + c_{1q}, \quad i,j = 1,2. \tag{6.21}$$

6.2. Spherical Particle

As presented in Section 5.1, the absolute values $\left|u'_{1p}\right|$, $\left|\varepsilon'_{ijp}\right|$, $\left|\sigma'_{ijp}\right|$ ($i,j = 1,2,3$) are required to represent increasing functions of $r \in \langle 0, R_p \rangle$. These increasing functions exhibit maximum values on the particle-matrix boundary, i.e., for $r = R_p$ (see Figure 2.1a). Due to $\varepsilon'_{ijp} \propto r^{\lambda_p - 1}$, $\sigma'_{ijp} \propto r^{\lambda_p - 1}$ ($i,j = 1,2,3$; see Equations (3.19)–(3.23), (3.36)–(3.39)), the exponent λ_p is required to be $\lambda_p > 1$, otherwise $(\varepsilon_{ijp})_{r\to 0} \to \pm\infty$, $(\sigma_{ijp})_{r\to 0} \to \pm\infty$. Due to $\lambda_{1p} > 3$, $\lambda_{2p} < -2$ (see Equations (6.4), (6.8)), we get $C_{1p} \neq 0$, $C_{2p} = 0$.

With regard to Equations (5.1), (6.7), the integration constant C_{1p} (see Equation (6.7)), which is determined by the boundary condition (5.1), has the form

$$C_{1p} = -\frac{p_p}{\xi_{1p}\xi_{3p}R_p^{\lambda_{1p}-1}}, \tag{6.22}$$

where λ_{1p}, ξ_{1p}, ξ_{3p} are derived by Equations (6.4), (6.9), (6.19), respectively. The radial stress p_p, which acts at the particle-matrix boundary, i.e., for $r = R_p$ (see Figure 2.1a), is determined in Section 6.5. With regard to Equations (6.10)–(6.17), (6.22), we get

$$u'_{1p} = -\frac{p_p r}{\xi_{3p}}\left(\frac{r}{R_p}\right)^{\lambda_{1p}-1}, \tag{6.23}$$

$$\varepsilon'_{11p} = -\frac{p_p \lambda_{1p}}{\xi_{3p}} \left(\frac{r}{R_p}\right)^{\lambda_{1p}-1}, \tag{6.24}$$

$$\varepsilon'_{22p} = \varepsilon'_{33p} = -\frac{p_p}{\xi_{3p}} \left(\frac{r}{R_p}\right)^{\lambda_{1p}-1}, \tag{6.25}$$

$$\varepsilon'_{12p} = -\frac{1}{\xi_{3p}} \frac{\partial p_p}{\partial \varphi} \left(\frac{r}{R_p}\right)^{\lambda_{1p}-1}, \tag{6.26}$$

$$\varepsilon'_{13p} = -\frac{1}{\xi_{3p}} \frac{\partial p_p}{\partial \nu} \left(\frac{r}{R_p}\right)^{\lambda_{1p}-1}, \tag{6.27}$$

$$\sigma'_{11p} = -p_p \left(\frac{r}{R_p}\right)^{\lambda_{1p}-1}, \tag{6.28}$$

$$\sigma'_{22p} = \sigma'_{33p} = -\frac{p_p \xi_{5p}}{\xi_{3p}} \left(\frac{r}{R_p}\right)^{\lambda_{1p}-1}, \tag{6.29}$$

$$\sigma'_{12p} = -\frac{1}{s_{44p}\xi_{3p}} \frac{\partial p_p}{\partial \varphi} \left(\frac{r}{R_p}\right)^{\lambda_{1p}-1}, \tag{6.30}$$

$$\sigma'_{13p} = -\frac{1}{s_{44p}\xi_{3p}} \frac{\partial p_p}{\partial \nu} \left(\frac{r}{R_p}\right)^{\lambda_{1p}-1}, \tag{6.31}$$

where s_{44p}, ξ_{5p} are given by Equations (3.35), (6.20), respectively.

6.3. Cell Matrix Around Particle

As presented in Section 5.2, the absolute values $|u'_{1m}|$, $\left|\varepsilon'_{ijm}\right|$, $\left|\sigma'_{ijm}\right|$ $(i, j - 1,2,3)$ are required to represent decreasing functions of $r \in \langle R_p, r_c \rangle$ (see Equations (3.19)–(3.23), (3.36)–(3.39)). These decreasing functions exhibit maximum values on the particle-matrix boundary, i.e., for $r = R_p$. These decreasing courses are ensured by the integration constants $C_{1m} \neq 0$, $C_{2m} \neq 0$, which are determined by the mandatory boundary conditions (5.3), (5.4). With regard to Equations (5.3), (5.4), (6.7)–(6.9), (6.14), we get

$$C_{im} = -\frac{1}{\xi_{im} r_c^{\lambda_{im}-1}} \left(\frac{p_p}{\zeta_{im}} + \frac{u'_{1c}\xi_{5-im}}{r_c \zeta_{2+im}}\right) = -\frac{1}{\xi_{im} r_c^{\lambda_{im}-1}} \left(\frac{p_p}{\zeta_{im}} + \frac{u_{1c0}\xi_{5-im}}{d \zeta_{2+im}}\right),$$
$$i = 1, 2, \tag{6.32}$$

where $u'_{1c}/r_c = u_{1c0}/d$ (see Equations (3.15)–(3.18), (5.4)); and r_c, ξ_{im} $(i=1,2)$ are given by Equations (3.15)–(3.18), (6.9), respectively. The radial stress p_p, which acts at the particle-matrix boundary, i.e., for $r = R_p$ (see Figure 2.1a), is determined in Section 6.5. The displacement u_{1c0} along the axis x_j $(j=1,2,3)$ on the cell surface *I*, *II* (see Equation (5.4)), i.e., for $r = d/2$ (see Figure 2.1), is determined in Section 6.6. The coefficients ζ_{im}, ζ_{2+im} $(i=1,2)$ have the forms

$$\zeta_{im} = \xi_{2+im}\left(\frac{R_p}{r_c}\right)^{\lambda_{im}-1} - \xi_{5-im}\left(\frac{R_p}{r_c}\right)^{\lambda_{3-im}-1}, \quad i=1,2, \tag{6.33}$$

$$\zeta_{2+im} = \xi_{2+im}\left(\frac{R_p}{r_c}\right)^{\lambda_{im}-\lambda_{3-im}} - \xi_{5-im}, \quad i=1,2. \tag{6.34}$$

where λ_{im} $(i=1,2)$, ξ_{3m}, ξ_{4m} are given by Equations (6.4), (6.19), respectively. With regard to Equations (6.10)–(6.17), (6.32), we get

$$u'_{1m} = -r\sum_{i=1}^{2}\left(\frac{p_p}{\zeta_{im}} + \frac{u_{1c0}\,\xi_{5-im}}{d\,\zeta_{2+im}}\right)\left(\frac{r}{r_c}\right)^{\lambda_{im}-1}, \tag{6.35}$$

$$\varepsilon'_{11m} = -\sum_{i=1}^{2}\lambda_{im}\left(\frac{p_p}{\zeta_{im}} + \frac{u_{1c0}\,\xi_{5-im}}{d\,\zeta_{2+im}}\right)\left(\frac{r}{r_c}\right)^{\lambda_{im}-1}, \tag{6.36}$$

$$\varepsilon'_{22m} = \varepsilon'_{33m} = -\sum_{i=1}^{2}\left(\frac{p_p}{\zeta_{im}} + \frac{u_{1c0}\,\xi_{5-im}}{d\,\zeta_{2+im}}\right)\left(\frac{r}{r_c}\right)^{\lambda_{im}-1}, \tag{6.37}$$

$$\begin{aligned}\varepsilon'_{12m} = &-\sum_{i=1}^{2}\frac{1-\lambda_{im}}{r_c}\left(\frac{p_p}{\zeta_{im}} + \frac{u_{1c0}\,\xi_{5-im}}{d\,\zeta_{2+im}}\right)\left(\frac{\partial r_c}{\partial \varphi}\right)\left(\frac{r}{r_c}\right)^{\lambda_{im}-1}\\ &-\sum_{i=1}^{2}\left[\frac{\partial}{\partial \varphi}\left(\frac{p_p}{\zeta_{im}}\right) - \frac{u_{1c0}\,\xi_{5-im}}{d\,\zeta_{2+im}^2}\frac{\partial \zeta_{2+im}}{\partial \varphi}\right]\left(\frac{r}{r_c}\right)^{\lambda_{im}-1},\end{aligned} \tag{6.38}$$

$$\begin{aligned}\varepsilon'_{13m} = &-\sum_{i=1}^{2}\frac{1-\lambda_{im}}{r_c}\left(\frac{p_p}{\zeta_{im}} + \frac{u_{1c0}\,\xi_{5-im}}{d\,\zeta_{2+im}}\right)\left(\frac{\partial r_c}{\partial \nu}\right)\left(\frac{r}{r_c}\right)^{\lambda_{im}-1}\\ &-\sum_{i=1}^{2}\left[\frac{\partial}{\partial \nu}\left(\frac{p_p}{\zeta_{im}}\right) - \frac{u_{1c0}\,\xi_{5-im}}{d\,\zeta_{2+im}^2}\frac{\partial \zeta_{2+im}}{\partial \nu}\right]\left(\frac{r}{r_c}\right)^{\lambda_{im}-1},\end{aligned} \tag{6.39}$$

$$\sigma'_{11m} = -\sum_{i=1}^{2} \xi_{2+im} \left(\frac{p_p}{\zeta_{im}} + \frac{u_{1c0}\xi_{5-im}}{d\,\zeta_{2+im}} \right) \left(\frac{r}{r_c} \right)^{\lambda_{im}-1}, \tag{6.40}$$

$$\sigma'_{22m} = \sigma'_{33m} = -\sum_{i=1}^{2} \xi_{4+im} \left(\frac{p_p}{\zeta_{im}} + \frac{u_{1c0}\xi_{5-im}}{d\,\zeta_{2+im}} \right) \left(\frac{r}{r_c} \right)^{\lambda_{im}-1}, \tag{6.41}$$

$$\begin{aligned} \sigma'_{12m} = & -\frac{1}{s_{44m}} \sum_{i=1}^{2} \frac{1-\lambda_{im}}{r_c} \left(\frac{p_p}{\zeta_{im}} + \frac{u_{1c0}\xi_{5-im}}{d\,\zeta_{2+im}} \right) \left(\frac{\partial r_c}{\partial \varphi} \right) \left(\frac{r}{r_c} \right)^{\lambda_{im}-1} \\ & -\frac{1}{s_{44m}} \sum_{i=1}^{2} \left[\frac{\partial}{\partial \varphi} \left(\frac{p_p}{\zeta_{im}} \right) - \frac{u_{1c0}\xi_{5-im}}{d\,\zeta_{2+im}^2} \frac{\partial \zeta_{2+im}}{\partial \varphi} \right] \left(\frac{r}{r_c} \right)^{\lambda_{im}-1} \end{aligned} \tag{6.42}$$

$$\begin{aligned} \sigma'_{13m} = & -\frac{1}{s_{44m}} \sum_{i=1}^{2} \frac{1-\lambda_{im}}{r_c} \left(\frac{p_p}{\zeta_{im}} + \frac{u_{1c0}\xi_{5-im}}{d\,\zeta_{2+im}} \right) \left(\frac{\partial r_c}{\partial \nu} \right) \left(\frac{r}{r_c} \right)^{\lambda_{im}-1} \\ & -\frac{1}{s_{44m}} \sum_{i=1}^{2} \left[\frac{\partial}{\partial \nu} \left(\frac{p_p}{\zeta_{im}} \right) - \frac{u_{1c0}\xi_{5-im}}{d\,\zeta_{2+im}^2} \frac{\partial \zeta_{2+im}}{\partial \nu} \right] \left(\frac{r}{r_c} \right)^{\lambda_{im}-1} \end{aligned} \tag{6.43}$$

where s_{44m} is given by Equation (3.35).

6.4. Cell Matrix Around Void

As presented in Section 5.3, the absolute values $|u'_{1v}|$, $\left|\varepsilon'_{ijv}\right|$, $\left|\sigma'_{ijv}\right|$ $(i, j = 1,2,3)$ are required to represent increasing functions of $r \in \langle R_v, r_c \rangle$ (see Equations (3.19)–(3.23), (3.36)–(3.39)). These increasing functions exhibit maximum values on the cell surface, i.e., for $r = r_c$ (see Equations (3.15)–(3.18)). These increasing courses are ensured by the integration constants $C_{1v} \neq 0$, $C_{2v} \neq 0$, which are determined by the mandatory boundary conditions (5.5), (5.6). With regard to Equations (5.5), (5.6), (6.7)–(6.9), (6.14), we get

$$C_{iv} = \frac{u'_{1c}\xi_{5-im}}{\xi_{im}\zeta_{iv} r_c^{\lambda_{im}}} = \frac{u_{1c0}\xi_{5-im}}{\xi_{im}\zeta_{iv}\, d\, r_c^{\lambda_{im}-1}}, \quad i = 1, 2, \tag{6.44}$$

where $u'_{1c}/r_c = u_{1c0}/d$ (see Equations (3.15)–(3.18), (5.4)); and r_c, ξ_{im}, ξ_{5-im} $(i = 1,2)$ are given by Equations (3.15)–(3.18), (6.9), (6.20), respectively. The displacement u_{1c0} along the axis x_j ($j = 1,2,3$) on the cell surface *I*, *II* (see Equation (5.4)), i.e., for $r = d/2$ (see Figure 2.1), is determined in Section 6.6. The coefficient ξ_{iv} ($i = 1,2$) has the form

$$\zeta_{iv} = \xi_{2+im}\left(\frac{R_v}{r_c}\right)^{\lambda_{im}-\lambda_{3-im}} - \xi_{5-im}, \quad i = 1,2. \tag{6.45}$$

where λ_{im} ($i = 1,2$), ξ_{3m}, ξ_{4m} are given by Equations (6.4), (6.19), respectively. With regard to Equations (6.10)–(6.17), (6.32), we get

$$u'_{1v} = \frac{r\,u_{1c0}}{d}\sum_{i=1}^{2}\frac{\xi_{5-im}}{\zeta_{iv}}\left(\frac{r}{r_c}\right)^{\lambda_{im}-1}, \tag{6.46}$$

$$\varepsilon'_{11v} = \frac{u_{1c0}}{d}\sum_{i=1}^{2}\frac{\lambda_{im}\xi_{5-im}}{\zeta_{iv}}\left(\frac{r}{r_c}\right)^{\lambda_{im}-1}, \tag{6.47}$$

$$\varepsilon'_{22v} = \varepsilon'_{33v} = \frac{u_{1c0}}{d}\sum_{i=1}^{2}\frac{\xi_{5-im}}{\zeta_{iv}}\left(\frac{r}{r_c}\right)^{\lambda_{im}-1}, \tag{6.48}$$

$$\varepsilon'_{12v} = \frac{u_{1c0}}{d}\sum_{i=1}^{2}\frac{\xi_{5-im}}{\zeta_{iv}}\left[\frac{1-\lambda_{im}}{r_c}\left(\frac{\partial r_c}{\partial \varphi}\right) - \frac{1}{\zeta_{iv}}\frac{\partial \zeta_{iv}}{\partial \varphi}\right]\left(\frac{r}{r_c}\right)^{\lambda_{im}-1}, \tag{6.49}$$

$$\varepsilon'_{13v} = \frac{u_{1c0}}{d}\sum_{i=1}^{2}\frac{\xi_{5-im}}{\zeta_{iv}}\left[\frac{1-\lambda_{im}}{r_c}\left(\frac{\partial r_c}{\partial \nu}\right) - \frac{1}{\zeta_{iv}}\frac{\partial \zeta_{iv}}{\partial \nu}\right]\left(\frac{r}{r_c}\right)^{\lambda_{im}-1}, \tag{6.50}$$

$$\sigma'_{11v} = \frac{u_{1c0}}{d}\sum_{i=1}^{2}\frac{\xi_{2+im}\xi_{5-im}}{\zeta_{iv}}\left(\frac{r}{r_c}\right)^{\lambda_{im}-1}, \tag{6.51}$$

$$\sigma'_{22v} = \sigma'_{33v} = \frac{u_{1c0}}{d}\sum_{i=1}^{2}\frac{\xi_{4+im}\xi_{5-im}}{\zeta_{iv}}\left(\frac{r}{r_c}\right)^{\lambda_{im}-1}, \tag{6.52}$$

$$\sigma'_{12v} = \frac{u_{1c0}}{d\,s_{44m}}\sum_{i=1}^{2}\frac{\xi_{5-im}}{\zeta_{iv}}\left[\frac{1-\lambda_{im}}{r_c}\left(\frac{\partial r_c}{\partial \varphi}\right) - \frac{1}{\zeta_{iv}}\frac{\partial \zeta_{iv}}{\partial \varphi}\right]\left(\frac{r}{r_c}\right)^{\lambda_{im}-1}, \tag{6.53}$$

$$\sigma'_{13v} = \frac{u_{1c0}}{d\,s_{44m}}\sum_{i=1}^{2}\frac{\xi_{5-im}}{\zeta_{iv}}\left[\frac{1-\lambda_{im}}{r_c}\left(\frac{\partial r_c}{\partial \nu}\right) - \frac{1}{\zeta_{iv}}\frac{\partial \zeta_{iv}}{\partial \nu}\right]\left(\frac{r}{r_c}\right)^{\lambda_{im}-1}, \tag{6.54}$$

where s_{44m} is given by Equation (3.35).

6.5. Radial Stress p_p

With regard to Equations (5.12), (6.25), (6.37), the radial stress p_p, which acts at the particle-matrix boundary, i.e., for $r = R_p$ (see Figure 2.1a), has the form

$$p_p = \zeta_{pm}\left(\frac{u'_{1c}\zeta_{6m}}{r_c} + \beta_m - \beta_p\right) = \zeta_{pm}\left(\frac{u_{1c0}\zeta_{6m}}{d} + \beta_m - \beta_p\right), \qquad (6.55)$$

where $u'_{1c}/r_c = u_{1c0}/d$ (see Equations (3.15)–(3.18), (5.4)). The coefficients ζ_{pm}, ζ_{5m}, ζ_{6m} are derived as

$$\zeta_{pm} = \frac{\xi_{3p}}{1 - \xi_{3p}\zeta_{5m}}, \qquad (6.56)$$

$$\zeta_{5m} = \frac{\left(\frac{R_p}{r_c}\right)^{\lambda_{1m}-1} - \left(\frac{R_p}{r_c}\right)^{\lambda_{2m}-1}}{\xi_{3m}\left(\frac{R_p}{r_c}\right)^{\lambda_{1m}-1} - \xi_{4m}\left(\frac{R_p}{r_c}\right)^{\lambda_{2m}-1}}, \qquad (6.57)$$

$$\zeta_{6m} = \frac{\left(\xi_{3m} + \xi_{4m}\right)\left(\frac{R_p}{r_c}\right)^{\lambda_{1m}-1}}{\xi_{3m}\left(\frac{R_p}{r_c}\right)^{\lambda_{1m}-\lambda_{2m}} - \xi_{4m}}, \qquad (6.58)$$

where r_c, λ_{im} $(i = 1,2)$, ξ_{3p}, ξ_{3m} ξ_{4m} and ξ_{5m} are given by Equations (3.15)–(3.18), (6.4), (6.19) and (6.20), respectively.

6.6. Displacement u_{1c0}

With regard to Equations (6.18), (6.22), (6.32), (6.44), the elastic energy density w_q $(q = p,m,v)$ has the form

$$w_p = \frac{r^{2\lambda_{1p}-2}}{\xi_{3p}^2}\left[\omega_{1p}\left(\frac{u_{1c0}}{d}\right)^2 + \frac{2\omega_{2p}u_{1c0}\left(\beta_m - \beta_p\right)}{d} + \omega_{3p}\left(\beta_m - \beta_p\right)^2\right], \qquad (6.59)$$

$$w_m = \omega_{1m}\left(\frac{u_{1c0}}{d}\right)^2 + \frac{\omega_{2m}u_{1c0}\left(\beta_m - \beta_p\right)}{d} + \omega_{3m}\left(\beta_m - \beta_p\right)^2, \qquad (6.60)$$

$$w_v = \left(\frac{u_{1c0}}{d}\right)^2 \sum_{i,j=1}^{2} \omega_{ijv} \left(\frac{r}{r_c}\right)^{\lambda_{im}+\lambda_{jm}-2}, \quad (6.61)$$

where the coefficients ω_{iq} ($i=1,2,3$; $q=p,m$), ω_{ijv} ($i, j=1,2$), along with the coefficients ζ_{1p}, ξ_{im} ($i=7,\ldots,36$), are derived as

$$\omega_{1p} = \zeta_{1p}\zeta_{6m}^2\zeta_{pm}^2 + \frac{1}{s_{44p}}\left\{\left[\frac{\partial(\zeta_{pm}\zeta_{6m})}{\partial\varphi}\right]^2 + \left[\frac{\partial(\zeta_{pm}\zeta_{6m})}{\partial\nu}\right]^2\right\}, \quad (6.62)$$

$$\omega_{2p} = \zeta_{1p}\zeta_{6m}\zeta_{pm}^2 + \frac{1}{s_{44p}}\left\{\frac{\partial\zeta_{pm}}{\partial\varphi}\left[\frac{\partial(\zeta_{pm}\zeta_{6m})}{\partial\varphi}\right]^2 + \frac{\partial\zeta_{pm}}{\partial\nu}\left[\frac{\partial(\zeta_{pm}\zeta_{6m})}{\partial\nu}\right]^2\right\}, \quad (6.63)$$

$$\omega_{3p} = \zeta_{1p}\zeta_{pm}^2 + \frac{1}{s_{44p}}\left[\left(\frac{\partial\zeta_{pm}}{\partial\varphi}\right)^2 + \left(\frac{\partial\zeta_{pm}}{\partial\nu}\right)^2\right], \quad (6.64)$$

$$\omega_{1m} = \sum_{i,j=1}^{2}\left(\xi_{6+i+2jm}\zeta_{4+i+2jm} + \frac{\zeta_{26+i+2jm}}{s_{44m}}\right)\left(\frac{r}{r_c}\right)^{\lambda_{im}+\lambda_{jm}-2}, \quad (6.65)$$

$$\omega_{2m} = \sum_{i,j=1}^{2}\left(\xi_{6+i+2jm}\zeta_{8+i+2jm} + \frac{\zeta_{30+i+2jm}}{s_{44m}}\right)\left(\frac{r}{r_c}\right)^{\lambda_{im}+\lambda_{jm}-2}, \quad (6.66)$$

$$\omega_{3m} = \sum_{i,j=1}^{2}\left(\xi_{6+i+2jm}\zeta_{12+i+2jm} + \frac{\zeta_{32+i+2jm}}{s_{44m}}\right)\left(\frac{r}{r_c}\right)^{\lambda_{im}+\lambda_{jm}-2}, \quad (6.67)$$

$$\begin{aligned}\omega_{ijv} = \frac{\xi_{5-im}\xi_{5-jm}}{\zeta_{iv}\zeta_{jv}}\Bigg\{\xi_{6+i+2jm} + \frac{1}{s_{44m}}&\left[\frac{1}{\zeta_{iv}}\frac{\partial\zeta_{iv}}{\partial\varphi} + r_c(\lambda_{im}-1)\frac{\partial r_c}{\partial\varphi}\right]\\ &\times\left[\frac{1}{\zeta_{jv}}\frac{\partial\zeta_{jv}}{\partial\varphi} + r_c(\lambda_{jm}-1)\frac{\partial r_c}{\partial\varphi}\right]\\ &\times\left[\frac{1}{\zeta_{iv}}\frac{\partial\zeta_{iv}}{\partial\nu} + r_c(\lambda_{im}-1)\frac{\partial r_c}{\partial\nu}\right]\\ &\times\left[\frac{1}{\zeta_{jv}}\frac{\partial\zeta_{jv}}{\partial\nu} + r_c(\lambda_{jm}-1)\frac{\partial r_c}{\partial\nu}\right]\Bigg\}, \quad (6.68)\end{aligned}$$

$$\zeta_{1p} = \lambda_{1p}\left[\frac{\lambda_{1p}(c_{1p}+c_{2p})}{2} - 2c_{2p}\right] + c_{1p}, \tag{6.69}$$

$$\zeta_{4+i+2jm} = \frac{\zeta_{6m}^2\zeta_{pm}^2}{\zeta_{im}\zeta_{jm}} + \frac{\xi_{5-im}\xi_{5-jm}}{\zeta_{2+im}\zeta_{2+jm}} + \zeta_{6m}\zeta_{pm}\left(\frac{\xi_{5-im}}{\zeta_{2+im}\zeta_{jm}} + \frac{\xi_{5-jm}}{\zeta_{im}\zeta_{2+jm}}\right),$$
$$i, j = 1,2, \tag{6.70}$$

$$\zeta_{8+i+2jm} = \frac{2\zeta_{6m}\zeta_{pm}^2}{\zeta_{im}\zeta_{jm}} + \zeta_{pm}\left(\frac{\xi_{5-im}}{\zeta_{2+im}\zeta_{jm}} + \frac{\xi_{5-jm}}{\zeta_{im}\zeta_{2+jm}}\right), \quad i, j = 1,2, \tag{6.71}$$

$$\zeta_{12+i+2jm} = \frac{\zeta_{pm}^2}{\zeta_{im}\zeta_{jm}}, \quad i, j = 1,2, \tag{6.72}$$

$$\zeta_{18+im} = \frac{\zeta_{6m}\zeta_{pm}}{\zeta_{im}} + \frac{\xi_{5-im}}{\zeta_{2+im}}, \quad i = 1,2, \tag{6.73}$$

$$\zeta_{19+2im} = \frac{\partial\zeta_{18+im}}{\partial\varphi} - \frac{d\zeta_{18+im}(\lambda_{im}-1)}{r_c}\frac{\partial f_c}{\partial\varphi}, \quad i = 1,2, \tag{6.74}$$

$$\zeta_{20+2im} = \frac{\partial}{\partial\varphi}\left(\frac{\zeta_{pm}}{\zeta_{im}}\right) - \frac{d\zeta_{pm}(\lambda_{im}-1)}{r_c\zeta_{im}}\frac{\partial f_c}{\partial\varphi}, \quad i = 1,2, \tag{6.75}$$

$$\zeta_{23+2im} = \frac{\partial\zeta_{18+im}}{\partial\nu} - \frac{d\zeta_{18+im}(\lambda_{im}-1)}{r_c}\frac{\partial f_c}{\partial\nu}, \quad i = 1,2, \tag{6.76}$$

$$\zeta_{24+2im} = \frac{\partial}{\partial\nu}\left(\frac{\zeta_{pm}}{\zeta_{im}}\right) - \frac{d\zeta_{pm}(\lambda_{im}-1)}{r_c\zeta_{im}}\frac{\partial f_c}{\partial\nu}, \quad i = 1,2, \tag{6.77}$$

$$\zeta_{26+2im} = \zeta_{19+2im}\zeta_{19+2jm} + \zeta_{23+2im}\zeta_{23+2jm}, \quad i = 1,2, \tag{6.78}$$

$$\zeta_{30+2im} = \zeta_{20+2im}\zeta_{19+2jm} + \zeta_{19+2im}\zeta_{20+2jm} + \zeta_{24+2im}\zeta_{23+2jm} + \zeta_{23+2im}\zeta_{24+2jm}, \quad i = 1,2, \tag{6.79}$$

$$\zeta_{32+2im} = \zeta_{20+2im}\zeta_{20+2jm} + \zeta_{24+2im}\zeta_{24+2jm}, \quad i = 1,2. \tag{6.80}$$

With regard to Equations (3.46)–(3.48), (6.59), the elastic energy W_q, which is accumulated in the spherical particle ($q = p$), has the form

$$W_p = \frac{8R_p^3}{\xi_{3p}^2 (2\lambda_{1p}+1)} \left[\Omega_{1p}\left(\frac{u_{1c0}}{d}\right)^2 + \frac{2u_{1c0}\Omega_{2p}(\beta_m - \beta_p)}{d} + \Omega_{3p}(\beta_m - \beta_p)^2\right], \tag{6.81}$$

where the coefficient Ω_{ip} ($i = 1,2,3$) is derived as

$$\Omega_{ip} = \int_0^{\pi/2}\int_0^{\pi/2} \omega_{ip}\, d\varphi\, d\nu, \quad i = 1,2,3. \tag{6.82}$$

With regard to Equations (3.46)–(3.48), (6.60), the elastic energy W_m, which is accumulated in the cell matrix I, has the form

$$W_m = 8d\left[u_{1c0}^2\,\Omega_{1m} + u_{1c0}\,d\,\Omega_{2m}(\beta_m - \beta_p) + d^2\,\Omega_{3m}(\beta_m - \beta_p)^2\right], \tag{6.83}$$

where the coefficient Ω_{im} ($i = 1,2,3$) is derived as

$$\Omega_{1m} = \int_0^{\pi/2}\int_0^{\pi/2} \sum_{i,j=1}^{2} \frac{f_c^3}{\lambda_{im}+\lambda_{jm}+1}\left(\xi_{6+i+2jm}\zeta_{4+i+2jm} + \frac{\zeta_{26+i+2jm}}{s_{44m}}\right) \times \left[1 - \left(\frac{R_p}{r_c}\right)^{\lambda_{im}+\lambda_{jm}+1}\right] d\varphi\, d\nu, \tag{6.84}$$

$$\Omega_{2m} = \int_0^{\pi/2}\int_0^{\pi/2} \sum_{i,j=1}^{2} \frac{f_c^3}{\lambda_{im}+\lambda_{jm}+1}\left(\xi_{6+i+2jm}\zeta_{8+i+2jm} + \frac{\zeta_{30+i+2jm}}{s_{44m}}\right) \times \left[1 - \left(\frac{R_p}{r_c}\right)^{\lambda_{im}+\lambda_{jm}+1}\right] d\varphi\, d\nu, \tag{6.85}$$

$$\Omega_{3m}=\int_0^{\pi/2}\int_0^{\pi/2}\sum_{i,j=1}^{2}\frac{f_c^3}{\lambda_{im}+\lambda_{jm}+1}\left(\xi_{6+i+2jm}\zeta_{12+i+2jm}+\frac{\zeta_{32+i+2jm}}{s_{44m}}\right)$$
$$\times\left[1-\left(\frac{R_p}{r_c}\right)^{\lambda_{im}+\lambda_{jm}+1}\right]d\varphi d\nu. \quad (6.86)$$

With regard to Equations (3.46)–(3.48), (6.61), the elastic energy W_m, which is accumulated in the cell matrix *II*, has the form

$$W_v=8\,d\,u_{1c0}^2\,\Omega_v, \quad (6.87)$$

where the coefficient Ω_v is derived as

$$\Omega_v=\int_0^{\pi/2}\int_0^{\pi/2}\sum_{i,j=1}^{2}\frac{\omega_{ijv}f_c^3}{\lambda_{im}+\lambda_{jm}+1}\left[1-\left(\frac{R_v}{r_c}\right)^{\lambda_{im}+\lambda_{jm}+1}\right]d\varphi d\nu. \quad (6.88)$$

With regard to Equations (5.13), (6.81), (6.83), (6.88), the displacement u_{1c0} along the axis x_i ($i=1,2,3$) on the cubic cell surface *I, II*, i.e., for $r=d/2$ (see Figure 2.1a), has the form

$$u_{1c0}=\frac{d\,\zeta_1\,(\beta_m-\beta_p)}{\zeta_2}, \quad (6.89)$$

where the coefficients ζ_1, ζ_2 are derived as

$$\zeta_1=\frac{\Omega_{2m}}{2}+\frac{\Omega_{2p}}{\xi_{3p}^2\,(2\lambda_{1p}+1)}\left(\frac{R_p}{d}\right)^3, \quad (6.90)$$

$$\zeta_2=\Omega_v-\left[\Omega_{1m}+\frac{\Omega_{1p}}{\xi_{3p}^2\,(2\lambda_{1p}+1)}\left(\frac{R_p}{d}\right)^3\right]. \quad (6.91)$$

Chapter 7

Analytical Model 2

The mathematical procedure in Section 7.1, which is applied to the differential equation (3.41), (3.42), results in the differential equation (7.3) with a non-zero right-hand side. Consequently, Wronskian's method is applied to Equation (7.3) [25]. The solution of Equation (7.3) includes the integration constants C_1, C_2, C_3 (see Equation (7.4)).

With regard to Equations (7.11)–(7.18), ε'_{ijq}, σ'_{ijq} (see Equations (3.19)–(3.23), (3.36)–(3.39)) are functions of $f_1 = \ln r$, $f_1 = r^{c_{3q}-1}$, $f_3 = 1/r$, where $c_{3q} < 0$ (see Equation (3.40)), and then we get $\left(\varepsilon'_{ijq}\right)_{r\to 0} \to \pm\infty$, $\left(\sigma'_{ijq}\right)_{r\to 0} \to \pm\infty$. Consequently, we get $C_{1p} = C_{2p} = C_{3p} = 0$, and the thermal and phase-transformation-induced stresses and strains in the spherical particle are determined in Section 6.2.

In case of the cell matrix *I, II* (see Figure 2.1a), the following solutions for the displacement $u'_{1q} = u'_{1q}(r, \varphi, \nu)$ $(q = m, v)$ are obtained:

1. the solutions (7.37), (7.51), which consider the integration constants C_{1q}, C_{2q} $(q = m, v)$, where $C_{3q} = 0$ (see Equations (7.32), (7.33), (7.46), (7.47)). These integration constants are determined by the mathematical boundary conditions (5.3)–(5.6).

2. the solutions (7.98), (7.111), which consider the integration constants C_{1q}, C_{3q} $(q = m, v)$, where $C_{2q} = 0$ (see Equations (7.94), (7.95), (7.107), (7.108)). These integration constants are determined by the mathematical boundary conditions (5.3)–(5.6).

3. the solutions (7.146), (7.159), which consider the integration constants C_{2q}, C_{3q} $(q = m,v)$, where $C_{1q} = 0$ (see Equations (7.142), (7.143), (7.155), (7.156)). These integration constants are determined by the mathematical boundary conditions (5.3)–(5.6).

4. the solution (7.195), (7.209), which consider the integration constants C_{1q}, C_{2q}, C_{3q} $(q = m,v)$ (see Equations (7.187), (7.204)). These integration constants are determined by the mathematical boundary conditions (5.3)–(5.8).

7.1. Mathematical Procedure 2

Performing $\partial^2 \mathrm{Eq.}(3.42)/\partial r^2$, the differential equation (3.42) is transformed to the form

$$r \frac{\partial^3 U_1'}{\partial r^3} + (2 - c_3) \frac{\partial^2 U_1'}{\partial r^2} = 0, \tag{7.1}$$

where the coefficient $c_3 < 0$ and the function $U_1' = U_1'(r, \varphi, \nu)$ are given by Equations (3.40) and (3.43), respectively. If U_1' is assumed in the form $U_1' = r^{\lambda}$, then the solution U_1' of Equation (7.1) has the form

$$U_1' = \frac{\partial^2 u_1'}{\partial \varphi^2} + \frac{\partial^2 u_1'}{\partial \nu^2} = C_1 r + C_2 r^{c_3} + C_3. \tag{7.2}$$

where the integration constant C_i $(i = 1,2,3)$ is determined by the boundary conditions, which are defined in Chapter 5.

Substituting Equation (6.2) to Equation (3.41), the differential equation (3.41) is transformed to the form

$$\frac{\partial^2 u_1'}{\partial r^2} + \frac{2}{r} \frac{\partial u_1'}{\partial r} - \frac{2u_1'}{r^2} = -\frac{1}{s_{44}(c_1 + c_2)} \left(\frac{C_1}{r} + C_2 r^{c_3 - 2} + \frac{C_3}{r^2} \right), \tag{7.3}$$

where s_{44} and c_1, c_2 are given by Equations (3.35) and (3.40), respectively.

Equation (7.3) with the variable r represents a linear differential equation of the second order with respect to u_1'. Due to the non-zero right-hand side, Wronskian's method is applied to Equation (7.3) [25]. Consequently, the radial displacement $u_{1q}' = u_{1q}'(r, \varphi, \nu)$ $(q = m,v)$ has the form

$$u'_{1q} = \sum_{i=1}^{2} C_{iq}\,\xi_{iq}\,u_{irq}, \quad q = m, v. \tag{7.4}$$

The subscripts $q = m$ and $q = v$ are related to the cell matrix *I* and *II* (see Figure 2.1a), where $r \in \langle R_p, r_c \rangle$ and $r \in \langle R_v, r_c \rangle$, respectively. The function $u_{1rq} = u_{1rq}(r)$, $u_{2rq} = u_{2rq}(r)$ ($q = m,v$), along with the constant function u_{3rq}, are derived as

$$u_{1rq} = r\left(\frac{1}{3} - \ln r\right), \tag{7.5}$$

$$u_{2rq} = r^{c_{3q}}, \quad c_{3q} < 0, \tag{7.6}$$

$$u_{3rq} = 1, \tag{7.7}$$

where $u_{1rq} = u_{1rq}(r)$, $u_{2rq} = u_{2rq}(r)$ are decreasing functions of the variable r due to $\partial u_{1rq}/\partial r = -\left[(2/3) + \ln r\right] < 0$ and $c_{3q} < 0$ (see Equation (3.40)). With regard to Equation (3.40), the coefficient ξ_{iq} ($i = 1,2,3$; $q = m,v$) has the form

$$\xi_{1q} = \frac{1}{3s_{44q}\left(c_{1q} + c_{2q}\right)} = \frac{1 - 2\mu_q}{6\left(1 - \mu_q\right)}, \tag{7.8}$$

$$\xi_{2q} = \xi_{1q}\left(\frac{1}{c_{3q} + 2} - \frac{1}{c_{3q} - 1}\right) = \frac{1}{4\left(1 - \mu_q\right)\left(4\mu_q - 5\right)}, \tag{7.9}$$

$$\xi_{3q} = \frac{3\xi_{1q}}{2} = \frac{1 - 2\mu_q}{4\left(1 - \mu_q\right)}. \tag{7.10}$$

With regard to Equations (3.19)–(3.23), (3.36)–(3.39), (3.45), (7.4)–(7.7), we get

$$\varepsilon'_{11q} = -C_{1q}\,\xi_{1q}\left(\frac{2}{3} + \ln r\right) + C_{2q}\,\xi_{2q}\,c_{3q}\,r^{c_{3q}-1}, \tag{7.11}$$

$$\varepsilon'_{22q} = \varepsilon'_{33q} = C_{1q}\,\xi_{1q}\left(\frac{1}{3} - \ln r\right) + C_{2q}\,\xi_{2q}\,r^{c_{3q}-1} + \frac{C_{3q}\,\xi_{3q}}{r}, \tag{7.12}$$

$$\varepsilon'_{12q} = \xi_{1q}\left(\frac{1}{3} - \ln r\right)\frac{\partial C_{1q}}{\partial \varphi} + \xi_{2q}\,r^{c_{3q}-1}\,\frac{\partial C_{2q}}{\partial \varphi} + \frac{\xi_{3q}}{r}\,\frac{\partial C_{3q}}{\partial \varphi}, \tag{7.13}$$

$$\varepsilon'_{13q} = \xi_{1q}\left(\frac{1}{3} - \ln r\right)\frac{\partial C_{1q}}{\partial \nu} + \xi_{2q}\, r^{c_{3q}-1}\frac{\partial C_{2q}}{\partial \nu} + \frac{\xi_{3q}}{r}\frac{\partial C_{3q}}{\partial \nu}, \tag{7.14}$$

$$\sigma'_{11q} = -C_{1q}\xi_{1q}\left[\frac{2\left(c_{1q} + 2c_{2q}\right)}{3} + \left(c_{1q} - c_{2q}\right)\ln r\right] + C_{2q}\xi_{2q}\left[\left(c_{1q} + c_{2q}\right)c_{3q} - 2c_{2q}\right] r^{c_{3q}-1} - \frac{2C_{3q}\xi_{3q}c_{2q}}{r}, \tag{7.15}$$

$$\sigma'_{22q} = \sigma'_{33q} = C_{1q}\xi_{1q}\left[\frac{c_{1q} + 2c_{2q}}{3} - \left(c_{1q} - c_{2q}\right)\ln r\right] + C_{2q}\xi_{2q}\left(c_{1q} - c_{2q}c_{3q}\right) r^{c_{3q}-1} + \frac{C_{3q}\xi_{3q}c_{1q}}{r}, \tag{7.16}$$

$$\sigma'_{12q} = \frac{1}{s_{44q}}\left[\xi_{1q}\left(\frac{1}{3} - \ln r\right)\frac{\partial C_{1q}}{\partial \varphi} + \xi_{2q}\, r^{c_{3q}-1}\frac{\partial C_{2q}}{\partial \varphi} + \frac{\xi_{3q}}{r}\frac{\partial C_{3q}}{\partial \varphi}\right], \tag{7.17}$$

$$\sigma'_{13q} = \frac{1}{s_{44q}}\left[\xi_{1q}\left(\frac{1}{3} - \ln r\right)\frac{\partial C_{1q}}{\partial \nu} + \xi_{2q}\, r^{c_{3q}-1}\frac{\partial C_{2q}}{\partial \nu} + \frac{\xi_{3q}}{r}\frac{\partial C_{3q}}{\partial \nu}\right], \tag{7.18}$$

$$w_q = \sum_{i,j=1;\, i\leq j}^{3} \xi_{iq}\xi_{jq}\left[C_{iq}C_{jq}u_{ijq}^{(r)} + \frac{u_{ijq}^{(\varphi\nu)}}{s_{44m}}\left(\frac{\partial C_i}{\partial \varphi}\frac{\partial C_j}{\partial \varphi} + \frac{\partial C_i}{\partial \nu}\frac{\partial C_j}{\partial \nu}\right)\right], \tag{7.19}$$

where the functions $u_{ijq}^{(r)} = u_{ijq}^{(r)}(r)$, $u_{ijq}^{(\varphi\nu)} = u_{ijq}^{(\varphi\nu)}(r)$ $(i, j = 1,2,3;\ i \leq j)$ are derived as

$$u_{11q}^{(r)} = \frac{c_{2q} - c_{1q}}{2}\ln^2 r + \frac{2\left(c_{2q} - c_{1q}\right)}{3}\ln r + \frac{7c_{1q} + 2c_{2q}}{9}, \tag{7.20}$$

$$u_{12q}^{(r)} = c_{3q}\left(c_{1q} - c_{2q}\right) r^{c_{3q}-1}\ln r + 2\left[c_{1q} - \frac{c_{3q}\left(2c_{1q} + c_{2q}\right)}{3}\right] r^{c_{3q}-1}, \tag{7.21}$$

$$u_{13q}^{(r)} = \frac{2c_{1q}}{r}, \tag{7.22}$$

$$u_{22q}^{(r)} = \left[\frac{c_{3q}^2 (c_{1q} + c_{2q})}{2} + c_{1q} (1 - 2c_{3q}) \right] r^{2(c_{3q}-1)}, \tag{7.23}$$

$$u_{23q}^{(r)} = 0, \tag{7.24}$$

$$u_{33q}^{(r)} = \frac{c_{1q}}{r^2}, \tag{7.25}$$

$$u_{11q}^{(\varphi\nu)} = \ln^2 r - \frac{2}{3} \ln r + \frac{1}{9}, \tag{7.26}$$

$$u_{12q}^{(\varphi\nu)} = \frac{2}{3} r^{c_{3q}-1} - 2r^{c_{3q}-1} \ln r, \tag{7.27}$$

$$u_{13q}^{(\varphi\nu)} = \frac{2}{3r} - \frac{2\ln r}{r}, \tag{7.28}$$

$$u_{22q}^{(\varphi\nu)} = r^{2(c_{3q}-1)}, \tag{7.29}$$

$$u_{23q}^{(\varphi\nu)} = r^{c_{3q}-2}, \tag{7.30}$$

$$u_{33q}^{(\varphi\nu)} - \frac{1}{r^2}. \tag{7.31}$$

7.2. Integration Constants $C_{1q} \neq 0, C_{2q} \neq 0, C_{3q} = 0$

As presented above (see page 53), the integration constant C_{1q}, C_{2q}, C_{3q} $(q = m, v)$ (see Equation (7.4)) are determined with respect to the following conditions

- $C_{1q} \neq 0, C_{2q} \neq 0, C_{3q} = 0,$
- $C_{1q} \neq 0, C_{3q} \neq 0, C_{2q} = 0,$
- $C_{1q} = 0, C_{2q} \neq 0, C_{3q} \neq 0,$
- $C_{1q} \neq 0, C_{2q} \neq 0, C_{3q} \neq 0.$

7.2.1. Cell Matrix around Particle

As presented above (see page 55), $u_{1rm} = u_{1rm}(r)$, $u_{2rm} = u_{2rm}(r)$ (see Equations (7.5), (7.6)) are decreasing functions of the variable r due to $\partial u_{1rm}/\partial r = -[(2/3)+\ln r] < 0$ and $c_{3m} < 0$ (see Equation (3.40)), respectively.

As presented in Section 5.2, the absolute values $|u'_{1m}|$, $\left|\varepsilon'_{ijm}\right|$, $\left|\sigma'_{ijm}\right|$ $(i,j=1,2,3)$ are required to represent decreasing functions of $r \in \langle R_p, r_c\rangle$ (see Equations (3.19)–(3.23), (3.36)–(3.39)). These decreasing functions exhibit maximum values on the particle-matrix boundary, i.e., for $r = R_p$. These decreasing courses are ensured by the integration constants $C_{1m} \neq 0$, $C_{2m} \neq 0$, which are determined by the mandatory boundary conditions (5.3), (5.4), where $C_{3m} = 0$. With regard to Equations (5.3), (5.4), (7.4)–(7.6), (7.15), we get

$$C_{1m} = -\frac{1}{\xi_{1m}\zeta_m}\left(p_p + \frac{u'_{1c}\zeta_{1m}}{r_c}\right) = -\frac{1}{\xi_{1m}\zeta_m}\left(p_p + \frac{u_{1c0}\zeta_{1m}}{d}\right), \tag{7.32}$$

$$\begin{aligned} C_{2m} &= \frac{1}{\xi_{2m}\zeta_m r_c^{c_{3m}-1}}\left[p_p\left(\frac{1}{3}-\ln r_c\right) + \frac{u'_{1c}\zeta_{2m}}{r_c}\right] \\ &= \frac{1}{\xi_{2m}\zeta_m r_c^{c_{3m}-1}}\left[p_p\left(\frac{1}{3}-\ln r_c\right) + \frac{u_{1c0}\zeta_{2m}}{d}\right], \end{aligned} \tag{7.33}$$

where $u'_{1c}/r_c = u_{1c0}/d$ (see Equations (3.15)–(3.18), (5.4)); and r_c, ξ_{im} $(i=1,2)$ are given by Equations (3.15)–(3.18), (7.8), (7.9), respectively. The radial stress p_p, which acts at the particle-matrix boundary, i.e., for $r = R_p$ (see Figure 2.1a), is determined in Section 7.2.3. The displacement u_{1c0} along the axis x_j $(j=1,2,3)$ on the cell surface *I*, *II* (see Equation (5.4)), i.e., for $r = d/2$ (see Figure 2.1), is determined in Section 7.2.4. The coefficients ζ_{im} $(i=1,2)$, ζ_m have the forms

$$\zeta_{1m} = [(c_{1m}+c_{2m})c_{3m} - 2c_{2m}]\left(\frac{R_p}{r_c}\right)^{c_{3m}-1}, \tag{7.34}$$

$$\zeta_{2m} = -\left[\frac{2(c_{1m}+2c_{2m})}{3} + (c_{1m}-c_{2m})\ln R_p\right], \tag{7.35}$$

$$\zeta_m = \zeta_{2m} - \zeta_{1m}\left(\frac{1}{3}-\ln r_c\right), \tag{7.36}$$

where c_{im} $(i=1,2,3)$ is given by Equation (3.40). With regard to Equations (7.4)–(7.6), (7.11)–(7.18), (7.32), (7.33), we get

$$u'_{1m} = -\frac{r}{\zeta_m}\left\{p_p\left[\frac{1}{3}-\ln r-\left(\frac{1}{3}-\ln r_c\right)\left(\frac{r}{r_c}\right)^{c_{3m}-1}\right]\right.$$
$$\left.+\frac{u_{1c0}}{d}\left[\zeta_{1m}\left(\frac{1}{3}-\ln r\right)-\zeta_{2m}\left(\frac{r}{r_c}\right)^{c_{3m}-1}\right]\right\}, \quad (7.37)$$

$$\varepsilon'_{11m} = \frac{1}{\zeta_m}\left\{p_p\left[\frac{2}{3}+\ln r+c_{3m}\left(\frac{1}{3}-\ln r_c\right)\left(\frac{r}{r_c}\right)^{c_{3m}-1}\right]\right.$$
$$\left.+\frac{u_{1c0}}{d}\left[\zeta_{1m}\left(\frac{2}{3}+\ln r\right)+c_{3m}\,\zeta_{2m}\left(\frac{r}{r_c}\right)^{c_{3m}-1}\right]\right\}, \quad (7.38)$$

$$\varepsilon'_{22m} = \varepsilon'_{33m} = -\frac{1}{\zeta_m}\left\{p_p\left[\frac{1}{3}-\ln r-\left(\frac{1}{3}-\ln r_c\right)\left(\frac{r}{r_c}\right)^{c_{3m}-1}\right]\right.$$
$$\left.+\frac{u_{1c0}}{d}\left[\zeta_{1m}\left(\frac{1}{3}-\ln r\right)-\zeta_{2m}\left(\frac{r}{r_c}\right)^{c_{3m}-1}\right]\right\}, \quad (7.39)$$

$$\varepsilon'_{12m} = -\frac{1}{\zeta_m}\left(\left[\frac{1}{3}-\ln r-\left(\frac{1}{3}-\ln r_c\right)\left(\frac{r}{r_c}\right)^{c_{3m}-1}\right]\left(\frac{\partial p_p}{\partial\varphi}-\frac{p_p}{\zeta_m}\frac{\partial\zeta_m}{\partial\varphi}\right)\right.$$
$$+\frac{p_p}{r_c}\left[1-(1-c_{3m})\left(\frac{1}{3}-\ln r_c\right)\right]\left(\frac{r}{r_c}\right)^{c_{3m}-1}\frac{\partial r_c}{\partial\varphi}$$
$$-\frac{u_{1c0}}{d\,\zeta_m}\left[\zeta_{1m}\left(\frac{1}{3}-\ln r\right)-\zeta_{2m}\left(\frac{r}{r_c}\right)^{c_{3m}-1}\right]\frac{\partial\zeta_m}{\partial\varphi}$$
$$+\frac{u_{1c0}}{d}\left[\left(\frac{1}{3}-\ln r\right)\frac{\partial\zeta_{1m}}{\partial\varphi}-\left(\frac{r}{r_c}\right)^{c_{3m}-1}\frac{\partial\zeta_{2m}}{\partial\varphi}\right.$$
$$\left.\left.-\frac{\zeta_{2m}(1-c_{3m})}{r_c}\left(\frac{r}{r_c}\right)^{c_{3m}-1}\frac{\partial r_c}{\partial\varphi}\right]\right), \quad (7.40)$$

$$\varepsilon'_{13m} = -\frac{1}{\zeta_m}\left(\left[\frac{1}{3} - \ln r - \left(\frac{1}{3} - \ln r_c\right)\left(\frac{r}{r_c}\right)^{c_{3m}-1}\right]\left(\frac{\partial p_p}{\partial \nu} - \frac{p_p}{\zeta_m}\frac{\partial \zeta_m}{\partial \nu}\right)\right.$$
$$+ \frac{p_p}{r_c}\left[1 - (1 - c_{3m})\left(\frac{1}{3} - \ln r_c\right)\right]\left(\frac{r}{r_c}\right)^{c_{3m}-1}\frac{\partial r_c}{\partial \nu}$$
$$- \frac{u_{1c0}}{d\,\zeta_m}\left[\zeta_{1m}\left(\frac{1}{3} - \ln r\right) - \zeta_{2m}\left(\frac{r}{r_c}\right)^{c_{3m}-1}\right]\frac{\partial \zeta_m}{\partial \nu}$$
$$+ \frac{u_{1c0}}{d}\left[\left(\frac{1}{3} - \ln r\right)\frac{\partial \zeta_{1m}}{\partial \nu} - \left(\frac{r}{r_c}\right)^{c_{3m}-1}\frac{\partial \zeta_{2m}}{\partial \nu}\right.$$
$$\left.\left. - \frac{\zeta_{2m}(1 - c_{3m})}{r_c}\left(\frac{r}{r_c}\right)^{c_{3m}-1}\frac{\partial r_c}{\partial \nu}\right]\right), \quad (7.41)$$

$$\sigma'_{11m} = \frac{1}{\zeta_m}\left(p_p\left\{\frac{2(c_{1m} + 2c_{2m})}{3} + (c_{1m} - c_{2m})\ln r\right.\right.$$
$$\left. + [(c_{1m} + c_{2m})c_{3m} - 2c_{2m}]\left(\frac{1}{3} - \ln r_c\right)\left(\frac{r}{r_c}\right)^{c_{3m}-1}\right\}$$
$$+ \frac{u_{1c0}}{d}\left\{\zeta_{1m}\left[\frac{2(c_{1m} + 2c_{2m})}{3} + (c_{1m} - c_{2m})\ln r\right]\right.$$
$$\left.\left. + \zeta_{2m}[(c_{1m} + c_{2m})c_{3m} - 2c_{2m}]\left(\frac{r}{r_c}\right)^{c_{3m}-1}\right\}\right), \quad (7.42)$$

$$\sigma'_{22m} = \sigma'_{33m} = -\frac{1}{\zeta_m}\left(p_p\left[\frac{c_{1m} + 2c_{2m}}{3} - (c_{1m} - c_{2m})\ln r\right.\right.$$
$$\left. - (c_{1m} - c_{2m}c_{3m})\left(\frac{1}{3} - \ln r_c\right)\left(\frac{r}{r_c}\right)^{c_{3m}-1}\right]$$
$$+ \frac{u_{1c0}}{d}\left\{\zeta_{1m}\left[\frac{c_{1m} + 2c_{2m}}{3} - (c_{1m} - c_{2m})\ln r\right]\right.$$
$$\left.\left. - \zeta_{2m}(c_{1m} - c_{2m}c_{3m})\left(\frac{r}{r_c}\right)^{c_{3m}-1}\right\}\right), \quad (7.43)$$

$$\sigma'_{12m} = -\frac{1}{s_{44m}\zeta_m}\left(\left[\frac{1}{3}-\ln r-\left(\frac{1}{3}-\ln r_c\right)\left(\frac{r}{r_c}\right)^{c_{3m}-1}\right]\right.$$
$$\times\left(\frac{\partial p_p}{\partial\varphi}-\frac{p_p}{\zeta_m}\frac{\partial\zeta_m}{\partial\varphi}\right)$$
$$+\frac{p_p}{r_c}\left[1-(1-c_{3m})\left(\frac{1}{3}-\ln r_c\right)\right]\left(\frac{r}{r_c}\right)^{c_{3m}-1}\frac{\partial r_c}{\partial\varphi}$$
$$-\frac{u_{1c0}}{d\zeta_m}\left[\zeta_{1m}\left(\frac{1}{3}-\ln r\right)-\zeta_{2m}\left(\frac{r}{r_c}\right)^{c_{3m}-1}\right]\frac{\partial\zeta_m}{\partial\varphi}$$
$$+\frac{u_{1c0}}{d}\left[\left(\frac{1}{3}-\ln r\right)\frac{\partial\zeta_{1m}}{\partial\varphi}-\left(\frac{r}{r_c}\right)^{c_{3m}-1}\frac{\partial\zeta_{2m}}{\partial\varphi}\right.$$
$$\left.\left.-\frac{\zeta_{2m}(1-c_{3m})}{r_c}\left(\frac{r}{r_c}\right)^{c_{3m}-1}\frac{\partial r_c}{\partial\varphi}\right]\right), \quad (7.44)$$

$$\sigma'_{13m} = -\frac{1}{s_{44m}\zeta_m}\left(\left[\frac{1}{3}-\ln r-\left(\frac{1}{3}-\ln r_c\right)\left(\frac{r}{r_c}\right)^{c_{3m}-1}\right]\right.$$
$$\times\left(\frac{\partial p_p}{\partial\nu}-\frac{p_p}{\zeta_m}\frac{\partial\zeta_m}{\partial\nu}\right)$$
$$+\frac{p_p}{r_c}\left[1-(1-c_{3m})\left(\frac{1}{3}-\ln r_c\right)\right]\left(\frac{r}{r_c}\right)^{c_{3m}-1}\frac{\partial r_c}{\partial\nu}$$
$$-\frac{u_{1c0}}{d\zeta_m}\left[\zeta_{1m}\left(\frac{1}{3}-\ln r\right)-\zeta_{2m}\left(\frac{r}{r_c}\right)^{c_{3m}-1}\right]\frac{\partial\zeta_m}{\partial\nu}$$
$$+\frac{u_{1c0}}{d}\left[\left(\frac{1}{3}-\ln r\right)\frac{\partial\zeta_{1m}}{\partial\nu}-\left(\frac{r}{r_c}\right)^{c_{3m}-1}\frac{\partial\zeta_{2m}}{\partial\nu}\right.$$
$$\left.\left.-\frac{\zeta_{2m}(1-c_{3m})}{r_c}\left(\frac{r}{r_c}\right)^{c_{3m}-1}\frac{\partial r_c}{\partial\nu}\right]\right), \quad (7.45)$$

where s_{44m} is given by Equation (3.35).

7.2.2. Cell Matrix around Void

As presented in Section 5.3, the absolute values $|u'_{1v}|$, $\left|\varepsilon'_{ijv}\right|$, $\left|\sigma'_{ijv}\right|$ $(i, j = 1,2,3)$ are required to represent increasing functions of $r \in \langle R_v, r_c \rangle$ (see Equations (3.19)–(3.23), (3.36)–(3.39)). These increasing functions exhibit maximum values on the cell surface, i.e., for $r = r_c$ (see Equations (3.15)–(3.18)). These increasing courses are ensured by the integration constants $C_{1v} \neq 0$, $C_{2v} \neq 0$, which are determined by the mandatory boundary conditions (5.5), (5.6), where $C_{3v} = 0$. With regard to Equations (5.5), (5.6), (7.4)–(7.6), (7.15), we get

$$C_{1v} = \frac{u'_{1c}\chi_{1v}}{\xi_{1v}\chi_v r_c} = \frac{u_{1c0}\chi_{1v}}{d\xi_{1v}\chi_v}, \tag{7.46}$$

$$C_{2v} = -\frac{u'_{1c}\chi_{2v}}{\xi_{2v}\chi_v r_c^{c_{3m}}} = -\frac{u_{1c0}\chi_{2v}}{d\xi_{2v}\chi_v r_c^{c_{3m}-1}}, \tag{7.47}$$

where $u'_{1c}/r_c = u_{1c0}/d$ (see Equations (3.15)–(3.18), (5.4)); and r_c, ξ_{iv} $(i = 1,2)$ are given by Equations (3.15)–(3.18), (7.8), (7.9), respectively. The displacement u_{1c0} along the axis x_j $(j = 1,2,3)$ on the cell surface *I*, *II* (see Equation (5.4)), i.e., for $r = d/2$ (see Figure 2.1), is determined in Section 7.2.4. The coefficients χ_{iv} $(i = 1,2)$, χ_v have the forms

$$\chi_{1v} = [(c_{1m} + c_{2m})c_{3m} - 2c_{2m}]\left(\frac{R_v}{r_c}\right)^{c_{3m}-1}, \tag{7.48}$$

$$\chi_{2v} = -\left[\frac{2(c_{1m} + 2c_{2m})}{3} + (c_{1m} - c_{2m})\ln R_v\right], \tag{7.49}$$

$$\chi_v = \zeta_{2m} - \zeta_{1m}\left(\frac{1}{3} - \ln r_c\right), \tag{7.50}$$

where c_{im} $(i = 1,2,3)$ is given by Equation (3.40). With regard to Equations (7.4), (7.5), (7.11)–(7.18), (7.46), (7.47), we get

$$u'_{1v} = \frac{r u_{1c0}}{d\chi_v}\left[\chi_{1v}\left(\frac{1}{3} - \ln r\right) - \chi_{2v}\left(\frac{r}{r_c}\right)^{c_{3m}-1}\right], \tag{7.51}$$

$$\varepsilon'_{11v} = -\frac{u_{1c0}}{d\chi_v}\left[\chi_{1v}\left(\frac{2}{3} + \ln r\right) + c_{3m}\chi_{2v}\left(\frac{r}{r_c}\right)^{c_{3m}-1}\right], \tag{7.52}$$

$$\varepsilon'_{22v} = \varepsilon'_{33v} = \frac{u_{1c0}}{d\chi_v}\left[\chi_{1v}\left(\frac{1}{3} - \ln r\right) - \chi_{2v}\left(\frac{r}{r_c}\right)^{c_{3m}-1}\right], \tag{7.53}$$

$$\begin{aligned}\varepsilon'_{12v} = -\frac{u_{1c0}}{d\chi_v}\Bigg\{&\frac{1}{\chi_v}\left[\chi_{1v}\left(\frac{1}{3} - \ln r\right) - \chi_{2v}\left(\frac{r}{r_c}\right)^{c_{3m}-1}\right]\frac{\partial\chi_v}{\partial\varphi}\\&-\left(\frac{1}{3} - \ln r\right)\frac{\partial\chi_{1v}}{\partial\varphi} + \left(\frac{r}{r_c}\right)^{c_{3m}-1}\frac{\partial\chi_{2v}}{\partial\varphi}\\&+\frac{\chi_{2v}(1-c_{3m})}{r_c}\left(\frac{r}{r_c}\right)^{c_{3m}-1}\frac{\partial r_c}{\partial\varphi}\Bigg\},\end{aligned} \tag{7.54}$$

$$\begin{aligned}\varepsilon'_{13v} = -\frac{u_{1c0}}{d\chi_v}\Bigg\{&\frac{1}{\chi_v}\left[\chi_{1v}\left(\frac{1}{3} - \ln r\right) - \chi_{2v}\left(\frac{r}{r_c}\right)^{c_{3m}-1}\right]\frac{\partial\chi_v}{\partial\nu}\\&-\left(\frac{1}{3} - \ln r\right)\frac{\partial\chi_{1v}}{\partial\nu} + \left(\frac{r}{r_c}\right)^{c_{3m}-1}\frac{\partial\chi_{2v}}{\partial\nu}\\&+\frac{\chi_{2v}(1-c_{3m})}{r_c}\left(\frac{r}{r_c}\right)^{c_{3m}-1}\frac{\partial r_c}{\partial\nu}\Bigg\},\end{aligned} \tag{7.55}$$

$$\begin{aligned}\sigma'_{11v} = -\frac{u_{1c0}}{d\chi_v}\Bigg\{&\chi_{1v}\left[\frac{2(c_{1m}+2c_{2m})}{3} + (c_{1m}-c_{2m})\ln r\right]\\&+\chi_{2v}\left[(c_{1m}+c_{2m})c_{3m} - 2c_{2m}\right]\left(\frac{r}{r_c}\right)^{c_{3m}-1}\Bigg\},\end{aligned} \tag{7.56}$$

$$\begin{aligned}\sigma'_{22v} = \sigma'_{33v} = \frac{u_{1c0}}{d\chi_v}\Bigg\{&\chi_{1v}\left[\frac{c_{1m}+2c_{2m}}{3} - (c_{1m}-c_{2m})\ln r\right]\\&-\chi_{2v}(c_{1m}-c_{2m}c_{3m})\left(\frac{r}{r_c}\right)^{c_{3m}-1}\Bigg\},\end{aligned} \tag{7.57}$$

$$\sigma'_{12v} = -\frac{u_{1c0}}{d\,s_{44m}\chi_v}\left\{\frac{1}{\chi_v}\left[\chi_{1v}\left(\frac{1}{3} - \ln r\right) - \chi_{2v}\left(\frac{r}{r_c}\right)^{c_{3m}-1}\right]\frac{\partial\chi_v}{\partial\varphi}\right.$$

$$- \left(\frac{1}{3} - \ln r \right) \frac{\partial \chi_{1v}}{\partial \varphi} + \left(\frac{r}{r_c} \right)^{c_{3m}-1} \frac{\partial \chi_{2v}}{\partial \varphi}$$
$$+ \frac{\chi_{2v} (1 - c_{3m})}{r_c} \left(\frac{r}{r_c} \right)^{c_{3m}-1} \frac{\partial r_c}{\partial \varphi} \Bigg\}, \tag{7.58}$$

$$\sigma'_{13v} = - \frac{u_{1c0}}{d\, s_{44m} \chi_v} \Bigg\{ \frac{1}{\chi_v} \left[\chi_{1v} \left(\frac{1}{3} - \ln r \right) - \chi_{2v} \left(\frac{r}{r_c} \right)^{c_{3m}-1} \right] \frac{\partial \chi_v}{\partial \nu}$$
$$- \left(\frac{1}{3} - \ln r \right) \frac{\partial \chi_{1v}}{\partial \nu} + \left(\frac{r}{r_c} \right)^{c_{3m}-1} \frac{\partial \chi_{2v}}{\partial \nu}$$
$$+ \frac{\chi_{2v} (1 - c_{3m})}{r_c} \left(\frac{r}{r_c} \right)^{c_{3m}-1} \frac{\partial r_c}{\partial \nu} \Bigg\}, \tag{7.59}$$

where s_{44m} is given by Equation (3.35).

7.2.3. Radial Stress p_p

With regard to Equations (5.12), (6.25), (7.39), the radial stress p_p, which acts at the particle-matrix boundary, i.e., for $r = R_p$ (see Figure 2.1a), has the form

$$p_p = \zeta_{pm} \left(\frac{u_{1c0} \zeta_{4m}}{d} + \beta_m - \beta_p \right), \tag{7.60}$$

where $u'_{1c}/r_c = u_{1c0}/d$ (see Equations (3.15)–(3.18), (5.4)). The coefficients ζ_{pm}, ζ_{3m}, ζ_{4m} are derived as

$$\zeta_{pm} = \frac{\xi_{3p}}{1 - \xi_{3p} \zeta_{3m}}, \tag{7.61}$$

$$\zeta_{3m} = \frac{1}{\zeta_m} \left[\zeta_{1m} \left(\frac{1}{3} - \ln R_p \right) - \left(\frac{1}{3} - \ln r_c \right) \left(\frac{R_p}{r_c} \right)^{c_{3m}-1} \right], \tag{7.62}$$

$$\zeta_{4m} = \frac{1}{\zeta_m} \left[\zeta_{1m} \left(\frac{1}{3} - \ln R_p \right) - \zeta_{2m} \left(\frac{R_p}{r_c} \right)^{c_{3m}-1} \right] \tag{7.63}$$

where r_c, c_{3m}, ξ_{3p}, ζ_{1m}, ζ_{2m}, ζ_m are given by Equations (3.15)–(3.18), (3.40), (6.19), (7.34)–(7.36), respectively.

7.2.4. Displacement u_{1c0}

With regard to Equations (7.32), (7.33), (7.46), (7.47), the integration constant C_{iq} ($i = 1,2$; $q = m,v$) has the form

$$C_{im} = \frac{1}{\xi_{im}}\left[\frac{u_{1c0}\zeta_{i1m}}{d} + \zeta_{i2m}(\beta_m - \beta_p)\right], \quad i = 1,2, \tag{7.64}$$

$$C_{iv} = \frac{u_{1c0}\zeta_{iv}}{d\xi_{iv}}, \quad i = 1,2, \tag{7.65}$$

where ξ_{im}, ξ_{iv} ($i = 1,2$) are given by Equations (7.8), (7.9), and β_q ($q = p,m$) is determined in Section 4.2. The coefficients ζ_{i1m}, ζ_{i2m}, ζ_{iv} ($i = 1,2$) are derived as

$$\zeta_{11m} = -\frac{\zeta_{1m} + \zeta_{pm}\zeta_{4m}}{\zeta_m}, \tag{7.66}$$

$$\zeta_{21m} = \frac{1}{\zeta_m r^{c_{3m}-1}}\left[\zeta_{pm}\zeta_{4m}\left(\frac{1}{3} - \ln r_c\right) + \zeta_{2m}\right], \tag{7.67}$$

$$\zeta_{12m} = -\frac{\zeta_{pm}}{\zeta_m}, \tag{7.68}$$

$$\zeta_{22m} = \frac{\zeta_{pm}}{\zeta_m r^{c_{3m}-1}}\left(\frac{1}{3} - \ln r_c\right), \tag{7.69}$$

$$\zeta_{1v} = \frac{\chi_{1v}}{\chi_v}, \tag{7.70}$$

$$\zeta_{2v} = -\frac{\chi_{2v}}{\chi_v r^{c_{3m}-1}}, \tag{7.71}$$

where r_c, ζ_m, ζ_{im}, χ_v, χ_{iv} ($i = 1,2$) and ζ_{pm}, ζ_{4m} are given by Equations (3.15)–(3.18), (7.34)–(7.36), (7.48)–(7.50) and (7.61), (7.63), respectively.

With regard to Equations (7.19)–(7.21), (7.23), (7.26), (7.27), (7.29), (7.64), (7.65), the elastic energy density w_q ($q = m,v$) has the form

$$w_m = \omega_{1m}^{(12)}\left(\frac{u_{1c0}}{d}\right)^2 + \frac{\omega_{2m}^{(12)} u_{1c0}(\beta_m - \beta_p)}{d} + \omega_{3m}^{(12)}(\beta_m - \beta_p)^2, \tag{7.72}$$

$$w_v = \omega_v^{(12)}\left(\frac{u_{1c0}}{d}\right)^2, \tag{7.73}$$

where the coefficients $\omega_{im}^{(12)}$ $(i=1,2,3)$, $\omega_{v}^{(12)}$ are derived as

$$\omega_{1m}^{(12)} = \zeta_{11m}^2 u_{11m}^{(r)} + \zeta_{21m}^2 u_{22m}^{(r)} + \zeta_{11m}\zeta_{21m} u_{12m}^{(r)} + \frac{u_{11m}^{(\varphi\nu)}}{s_{44m}}\left[\left(\frac{\partial\zeta_{11m}}{\partial\varphi}\right)^2 + \left(\frac{\partial\zeta_{11m}}{\partial\nu}\right)^2\right] + \frac{u_{22m}^{(\varphi\nu)}}{s_{44m}}\left[\left(\frac{\partial\zeta_{21m}}{\partial\varphi}\right)^2 + \left(\frac{\partial\zeta_{21m}}{\partial\nu}\right)^2\right] + \frac{u_{12m}^{(\varphi\nu)}}{s_{44m}}\left[\frac{\partial\zeta_{11m}}{\partial\varphi}\frac{\partial\zeta_{21m}}{\partial\varphi} + \frac{\partial\zeta_{11m}}{\partial\nu}\frac{\partial\zeta_{21m}}{\partial\nu}\right], \quad (7.74)$$

$$\omega_{2m}^{(12)} = 2\left[\zeta_{11m}\zeta_{12m} u_{11m}^{(r)} + \zeta_{21m}\zeta_{22m} u_{22m}^{(r)}\right] + \left(\zeta_{11m}\zeta_{22m} + \zeta_{12m}\zeta_{21m}\right) u_{12m}^{(r)} + \frac{2u_{11m}^{(\varphi\nu)}}{s_{44m}}\left(\frac{\partial\zeta_{11m}}{\partial\varphi}\frac{\partial\zeta_{12m}}{\partial\varphi} + \frac{\partial\zeta_{11m}}{\partial\nu}\frac{\partial\zeta_{12m}}{\partial\nu}\right) + \frac{2u_{22m}^{(\varphi\nu)}}{s_{44m}}\left(\frac{\partial\zeta_{21m}}{\partial\varphi}\frac{\partial\zeta_{22m}}{\partial\varphi} + \frac{\partial\zeta_{21m}}{\partial\nu}\frac{\partial\zeta_{22m}}{\partial\nu}\right) + \frac{u_{12m}^{(\varphi\nu)}}{s_{44m}}\left(\frac{\partial\zeta_{11m}}{\partial\varphi}\frac{\partial\zeta_{22m}}{\partial\varphi} + \frac{\partial\zeta_{12m}}{\partial\varphi}\frac{\partial\zeta_{21m}}{\partial\varphi} + \frac{\partial\zeta_{11m}}{\partial\nu}\frac{\partial\zeta_{22m}}{\partial\nu} + \frac{\partial\zeta_{12m}}{\partial\nu}\frac{\partial\zeta_{21m}}{\partial\nu}\right), \quad (7.75)$$

$$\omega_{3m}^{(12)} = \zeta_{12m}^2 u_{11m}^{(r)} + \zeta_{22m}^2 u_{22m}^{(r)} + \zeta_{12m}\zeta_{22m} u_{12m}^{(r)} + \frac{u_{11m}^{(\varphi\nu)}}{s_{44m}}\left[\left(\frac{\partial\zeta_{12m}}{\partial\varphi}\right)^2 + \left(\frac{\partial\zeta_{12m}}{\partial\nu}\right)^2\right] + \frac{u_{22m}^{(\varphi\nu)}}{s_{44m}}\left[\left(\frac{\partial\zeta_{22m}}{\partial\varphi}\right)^2 + \left(\frac{\partial\zeta_{22m}}{\partial\nu}\right)^2\right] + \frac{u_{12m}^{(\varphi\nu)}}{s_{44m}}\left[\frac{\partial\zeta_{12m}}{\partial\varphi}\frac{\partial\zeta_{22m}}{\partial\varphi} + \frac{\partial\zeta_{12m}}{\partial\nu}\frac{\partial\zeta_{22m}}{\partial\nu}\right], \quad (7.76)$$

$$\omega_{v}^{(12)} = \zeta_{1v}^2 u_{11m}^{(r)} + \zeta_{2v}^2 u_{22m}^{(r)} + \zeta_{1v}\zeta_{2v} u_{12m}^{(r)}$$

$$+\frac{u_{11m}^{(\varphi\nu)}}{s_{44m}}\left[\left(\frac{\partial\zeta_{1\nu}}{\partial\varphi}\right)^2+\left(\frac{\partial\zeta_{1\nu}}{\partial\nu}\right)^2\right]$$

$$+\frac{u_{22m}^{(\varphi\nu)}}{s_{44m}}\left[\left(\frac{\partial\zeta_{2\nu}}{\partial\varphi}\right)^2+\left(\frac{\partial\zeta_{2\nu}}{\partial\nu}\right)^2\right]$$

$$+\frac{u_{12m}^{(\varphi\nu)}}{s_{44m}}\left[\frac{\partial\zeta_{1\nu}}{\partial\varphi}\frac{\partial\zeta_{2\nu}}{\partial\varphi}+\frac{\partial\zeta_{1\nu}}{\partial\nu}\frac{\partial\zeta_{2\nu}}{\partial\nu}\right], \tag{7.77}$$

where s_{44m} is given by Equation (3.35).

With regard to Equations (3.47), (7.72), the elastic energy W_m, which is accumulated in the cell matrix I, has the form

$$W_m = 8\left[\Omega_{1m}^{(12)}\left(\frac{u_{1c0}}{d}\right)^2+\frac{\Omega_{2m}^{(12)}u_{1c0}\left(\beta_m-\beta_p\right)}{d}+\Omega_{3m}^{(12)}\left(\beta_m-\beta_p\right)^2\right]. \tag{7.78}$$

The coefficient $\Omega_{im}^{(12)}$ $(i=1,2,3)$ is derived as

$$\Omega_{1m}^{(12)} = \int_0^{\pi/2}\int_0^{\pi/2}\int_{R_p}^{r_c}\omega_{1m}^{(12)}r^2\,dr\,d\varphi\,d\nu$$

$$= \int_0^{\pi/2}\int_0^{\pi/2}\left\{\zeta_{11m}^2U_{11m}^{(r)}+\zeta_{21m}^2U_{22m}^{(r)}+\zeta_{11m}\zeta_{21m}U_{12m}^{(r)}\right.$$

$$+\frac{U_{11m}^{(\varphi\nu)}}{s_{44m}}\left[\left(\frac{\partial\zeta_{11m}}{\partial\varphi}\right)^2+\left(\frac{\partial\zeta_{11m}}{\partial\nu}\right)^2\right]$$

$$+\frac{U_{22m}^{(\varphi\nu)}}{s_{44m}}\left[\left(\frac{\partial\zeta_{21m}}{\partial\varphi}\right)^2+\left(\frac{\partial\zeta_{21m}}{\partial\nu}\right)^2\right]$$

$$\left.+\frac{U_{12m}^{(\varphi\nu)}}{s_{44m}}\left[\frac{\partial\zeta_{11m}}{\partial\varphi}\frac{\partial\zeta_{21m}}{\partial\varphi}+\frac{\partial\zeta_{11m}}{\partial\nu}\frac{\partial\zeta_{21m}}{\partial\nu}\right]\right\}d\varphi\,d\nu, \tag{7.79}$$

$$\Omega_{2m}^{(12)} = \int_0^{\pi/2}\int_0^{\pi/2}\int_{R_p}^{r_c}\omega_{2m}^{(12)}r^2\,dr\,d\varphi\,d\nu$$

$$
\begin{aligned}
&= \int_0^{\pi/2}\int_0^{\pi/2} \Bigg\{ 2\left[\zeta_{11m}\zeta_{12m}U_{11m}^{(r)} + \zeta_{21m}\zeta_{22m}U_{22m}^{(r)}\right] \\
&\qquad + \left(\zeta_{11m}\zeta_{22m} + \zeta_{12m}\zeta_{21m}\right)U_{12m}^{(r)} \\
&\qquad + \frac{2U_{11m}^{(\varphi\nu)}}{s_{44m}}\left(\frac{\partial\zeta_{11m}}{\partial\varphi}\frac{\partial\zeta_{12m}}{\partial\varphi} + \frac{\partial\zeta_{11m}}{\partial\nu}\frac{\partial\zeta_{12m}}{\partial\nu}\right) \\
&\qquad + \frac{2U_{22m}^{(\varphi\nu)}}{s_{44m}}\left(\frac{\partial\zeta_{21m}}{\partial\varphi}\frac{\partial\zeta_{22m}}{\partial\varphi} + \frac{\partial\zeta_{21m}}{\partial\nu}\frac{\partial\zeta_{22m}}{\partial\nu}\right) \\
&\qquad + \frac{U_{12m}^{(\varphi\nu)}}{s_{44m}}\left(\frac{\partial\zeta_{11m}}{\partial\varphi}\frac{\partial\zeta_{22m}}{\partial\varphi} + \frac{\partial\zeta_{12m}}{\partial\varphi}\frac{\partial\zeta_{21m}}{\partial\varphi}\right) \\
&\qquad + \frac{U_{12m}^{(\varphi\nu)}}{s_{44m}}\left(\frac{\partial\zeta_{11m}}{\partial\nu}\frac{\partial\zeta_{22m}}{\partial\nu} + \frac{\partial\zeta_{12m}}{\partial\nu}\frac{\partial\zeta_{21m}}{\partial\nu}\right)\Bigg\}\, d\varphi\, d\nu, \qquad (7.80)
\end{aligned}
$$

$$
\begin{aligned}
\Omega_{3m}^{(12)} &= \int_0^{\pi/2}\int_0^{\pi/2}\int_{R_p}^{r_c} \omega_{3m}^{(12)}\, r^2\, dr\, d\varphi\, d\nu \\
&= \int_0^{\pi/2}\int_0^{\pi/2} \Bigg\{ \zeta_{12m}^2 U_{11m}^{(r)} + \zeta_{22m}^2 U_{22m}^{(r)} + \zeta_{12m}\zeta_{22m}U_{12m}^{(r)} \\
&\qquad + \frac{U_{11m}^{(\varphi\nu)}}{s_{44m}}\left[\left(\frac{\partial\zeta_{12m}}{\partial\varphi}\right)^2 + \left(\frac{\partial\zeta_{12m}}{\partial\nu}\right)^2\right] \\
&\qquad + \frac{U_{22m}^{(\varphi\nu)}}{s_{44m}}\left[\left(\frac{\partial\zeta_{22m}}{\partial\varphi}\right)^2 + \left(\frac{\partial\zeta_{22m}}{\partial\nu}\right)^2\right] \\
&\qquad + \frac{U_{12m}^{(\varphi\nu)}}{s_{44m}}\left[\frac{\partial\zeta_{12m}}{\partial\varphi}\frac{\partial\zeta_{22m}}{\partial\varphi} + \frac{\partial\zeta_{12m}}{\partial\nu}\frac{\partial\zeta_{22m}}{\partial\nu}\right]\Bigg\}\, d\varphi\, d\nu, \qquad (7.81)
\end{aligned}
$$

where $U_{ijm}^{(x)}$ $(i, j = 1{,}2;\ x = r, \varphi\nu)$ has the form

$$
U_{ijm}^{(x)} = \int_{R_p}^{r_c} u_{ijm}^{(x)} r^2\, dr, \quad i, j = 1, 2, 3; \quad x = r,\ \varphi\nu. \qquad (7.82)
$$

With regard to Equations (7.20), (7.21), (7.23), (7.26), (7.27), (7.29), (7.82), we get

$$U_{11m}^{(r)} = \frac{c_{2m} - c_{1m}}{6} \left\{ r_c^3 \left[\left(\ln r_c - \frac{1}{3} \right)^2 + \frac{1}{9} \right] - R_p^3 \left[\left(\ln R_p - \frac{1}{3} \right)^2 + \frac{1}{9} \right] \right\} + \frac{2\,(c_{2m} - c_{1m})}{9} \left[r_c^3 \left(\ln r_c - \frac{1}{3} \right) - R_p^3 \left(\ln R_p - \frac{1}{3} \right) \right] + \frac{(7c_{1m} + 2\,c_{2m})\left(r_c^3 - R_p^3\right)}{27}, \tag{7.83}$$

$$U_{12m}^{(r)} = \frac{c_{3m}\,(c_{1m} - c_{2m})}{c_{3m} + 2} \left[r_c^{c_{3m}+2} \left(\ln r_c - \frac{1}{c_{3m}+2} \right) - R_p^{c_{3m}+2} \left(\ln R_p - \frac{1}{c_{3m}+2} \right) \right] + \frac{2}{c_{3m}+2} \left[c_{1m} - \frac{c_{3m}\,(2\,c_{1m} + c_{2m})}{3} \right] \left(r_c^{c_{3m}+2} - R_p^{c_{3m}+2} \right), \tag{7.84}$$

$$U_{22m}^{(r)} = \frac{1}{2\,c_{3m}+1} \left[\frac{c_{3m}^2\,(c_{1m} + c_{2m})}{2} + c_{1m}\,(1 - 2\,c_{3m}) \right] \left(r_c^{2\,c_{3m}+1} - R_p^{2\,c_{3m}+1} \right), \tag{7.85}$$

$$U_{11m}^{(\varphi\nu)} = \frac{r_c^3}{3} \left[(\ln r_c - 1) \left(\ln r_c - \frac{1}{3} \right) + \frac{2}{9} \right] - \frac{R_p^3}{3} \left[(\ln R_p - 1) \left(\ln R_p - \frac{1}{3} \right) + \frac{2}{9} \right], \tag{7.86}$$

$$U_{12m}^{(\varphi\nu)} = \frac{2}{c_{3m}+2} \left\{ r_c^{c_{3m}+2} \left[\frac{c_{3m}+5}{3\,(c_{3m}+2)} - \ln r_c \right] \right.$$

$$- R_p^{c_{3m}+2}\left[\frac{c_{3m}+5}{3\,(c_{3m}+2)} - \ln R_p\right]\Bigg\},$$

(7.87)

$$U_{22m}^{(\varphi\nu)} = \frac{r_c^{2\,c_{3m}+1} - R_p^{2\,c_{3m}+1}}{2\,c_{3m}+1}, \tag{7.88}$$

where r_c, c_{im} $(i = 1,2,3)$ are given by Equations (3.15)–(3.18), (3.40), respectively.

With regard to Equations (3.48), (7.73), the elastic energy W_v, which is accumulated in the cell matrix *II*, is derived as

$$W_v = 8\,\Omega_v^{(12)}\left(\frac{u_{1c0}}{d}\right)^2. \tag{7.89}$$

With regard to Equation (7.77), the coefficient $\Omega_v^{(12)}$ is derived as

$$\begin{aligned}\Omega_v^{(12)} &= \int_0^{\pi/2}\int_0^{\pi/2}\int_{R_v}^{r_c} \omega_v^{(12)}\, r^2\, dr\, d\varphi\, d\nu \\ &= \int_0^{\pi/2}\int_0^{\pi/2} \Bigg\{ \zeta_{1v}^2\, U_{11v}^{(r)} + \zeta_{2v}^2\, U_{22v}^{(r)} + \zeta_{1v}\,\zeta_{2v}\, U_{12v}^{(r)} \\ &\quad + \frac{U_{11v}^{(\varphi\nu)}}{s_{44m}}\left[\left(\frac{\partial\zeta_{1v}}{\partial\varphi}\right)^2 + \left(\frac{\partial\zeta_{1v}}{\partial\nu}\right)^2\right] \\ &\quad + \frac{U_{22v}^{(\varphi\nu)}}{s_{44m}}\left[\left(\frac{\partial\zeta_{2v}}{\partial\varphi}\right)^2 + \left(\frac{\partial\zeta_{2v}}{\partial\nu}\right)^2\right] \\ &\quad + \frac{U_{12v}^{(\varphi\nu)}}{s_{44m}}\left[\frac{\partial\zeta_{1v}}{\partial\varphi}\frac{\partial\zeta_{2v}}{\partial\varphi} + \frac{\partial\zeta_{1v}}{\partial\nu}\frac{\partial\zeta_{2v}}{\partial\nu}\right]\Bigg\}\, d\varphi\, d\nu,\end{aligned} \tag{7.90}$$

where $U_{ijv}^{(x)}$ $(i, j = 1,2,3;\ x = r, \varphi\nu)$ has the form

$$U_{ijv}^{(x)} = \int_{R_v}^{r_c} u_{ijm}^{(x)}\, r^2\, dr, \quad i, j = 1,2,3; \quad x = r,\ \varphi\nu. \tag{7.91}$$

With regard to Equations (7.82), (7.91), $U_{ijv}^{(x)}$ $(i,j=1,2;\ x=r,\ \varphi\nu)$ is given by Equations (7.83)–(7.88). In case of $U_{ijv}^{(x)}$, the radius R_p, which is included in Equations (7.83)–(7.88), is replaced by R_v.

With regard to Equations (5.13), (6.81), (7.78), (7.89), the displacement u_{1c0} along the axis x_i $(i=1,2,3)$ on the cubic cell surface *I, II*, i.e., for $r=d/2$ (see Figure 2.1a), is given by Equation (6.89). The coefficients ζ_1, ζ_2, which are included in Equation (6.89), are derived as

$$\zeta_1=\frac{\Omega_{2m}^{(12)}}{2}+\frac{\Omega_{2p}R_p^3}{\xi_{3p}^2\left(2\lambda_{1p}+1\right)},\tag{7.92}$$

$$\zeta_2=\Omega_v^{(12)}-\left[\Omega_{1m}^{(12)}+\frac{\Omega_{1p}R_p^3}{\xi_{3p}^2\left(2\lambda_{1p}+1\right)}\right],\tag{7.93}$$

where ξ_{3p}, Ω_{1p}, $\Omega_{1m}^{(12)}$, $\Omega_{2m}^{(12)}$, $\Omega_v^{(12)}$ are given by Equations (6.19), (7.79), (7.81), (7.90), respectively.

7.3. Integration Constants $C_{1q}\neq 0, C_{3q}\neq 0, C_{2q}=0$

7.3.1. Cell Matrix Around Particle

As presented above (see page 55), $u_{1rm}=u_{1rm}(r)$ is a decreasing function of the variable r due to $\partial u_{1rm}/\partial r=-\left[(2/3)+\ln r\right]<0$, and u_{3rm} is a constant function with respect to the variable r (see Equations (7.5), (7.7)).

As presented in Section 5.2, the absolute values $|u'_{1m}|$, $\left|\varepsilon'_{ijm}\right|$, $\left|\sigma'_{ijm}\right|$ $(i,j=1,2,3)$ are required to represent decreasing functions of $r\in\langle R_p,r_c\rangle$ (see Equations (3.19)–(3.23), (3.36) (3.39)). These decreasing functions exhibit maximum values on the particle-matrix boundary, i.e., for $r=R_p$. These decreasing courses are ensured by the integration constants $C_{1m}\neq 0$, $C_{3m}\neq 0$, which are determined by the mandatory boundary conditions (5.3), (5.4), where $C_{2m}=0$. With regard to Equations (5.3), (5.4), (7.4)–(7.6), (7.15), we get

$$\begin{aligned}C_{1m}&=-\frac{1}{\xi_{1m}\zeta_{1m}}\left(\frac{p_pR_p}{r_c}-\frac{2c_{2m}u'_{1c}}{r_c}\right)\\&=-\frac{1}{\xi_{1m}\zeta_{1m}}\left(\frac{p_pR_p}{r_c}-\frac{2c_{2m}u_{1c0}}{d}\right),\end{aligned}\tag{7.94}$$

$$C_{3m} = \frac{1}{\xi_{3m}\zeta_{1m}}\left[p_p R_p\left(\frac{1}{3} - \ln r_c\right) + \frac{u'_{1c}\zeta_{2m}}{r_c}\right]$$
$$= \frac{1}{\xi_{3m}\zeta_{1m}}\left[p_p R_p\left(\frac{1}{3} - \ln r_c\right) + \frac{u_{1c0}\zeta_{2m}}{d}\right], \quad (7.95)$$

where $u'_{1c}/r_c = u_{1c0}/d$ (see Equations (3.15)–(3.18), (5.4)); and r_c, ξ_{im} $(i=1,3)$ are given by Equations (3.15)–(3.18), (7.8), (7.10), respectively. The radial stress p_p, which acts at the particle-matrix boundary, i.e., for $r = R_p$ (see Figure 2.1a), is determined in Section 7.3.3. The displacement u_{1c0} along the axis x_j $(j=1,2,3)$ on the cell surface *I*, *II* (see Equation (5.4)), i.e., for $r = d/2$ (see Figure 2.1), is determined in Section 7.3.4. The coefficient ζ_{im} $(i=1,2)$ has the forms

$$\zeta_{1m} = \frac{R_p}{r_c}\left[\frac{2(c_{1m}+2c_{2m})}{3} + (c_{1m}-c_{2m})\ln R_p\right] + 2c_{2m}\left(\frac{1}{3} - \ln r_c\right), \quad (7.96)$$

$$\zeta_{2m} = R_p\left[\frac{2(c_{1m}+2c_{2m})}{3} + (c_{1m}-c_{2m})\ln R_p\right], \quad (7.97)$$

where c_{im} $(i=1,2)$ is given by Equation (3.40). With regard to Equations (7.4), (7.5), (7.7), (7.11)–(7.18), (7.94), (7.95), we get

$$u'_{1m} = -\frac{r}{\zeta_{1m}}\left\{p_p R_p\left[\frac{1}{r_c}\left(\frac{1}{3} - \ln r\right) - \frac{1}{r}\left(\frac{1}{3} - \ln r_c\right)\right]\right.$$
$$\left. - \frac{u_{1c0}}{d}\left[2c_{2m}\left(\frac{1}{3} - \ln r\right) + \frac{\zeta_{2m}}{r}\right]\right\}, \quad (7.98)$$

$$\varepsilon'_{11m} = \frac{1}{\zeta_{1m}}\left(\frac{p_p R_p}{r_c} - \frac{2c_{2m}u_{1c0}}{d}\right)\left(\frac{2}{3} + \ln r\right), \quad (7.99)$$

$$\varepsilon'_{22m} = \varepsilon'_{33m} = -\frac{1}{\zeta_{1m}}\left\{p_p R_p\left[\frac{1}{r_c}\left(\frac{1}{3} - \ln r\right) - \frac{1}{r}\left(\frac{1}{3} - \ln r_c\right)\right]\right.$$
$$\left. - \frac{u_{1c0}}{d}\left[2c_{2m}\left(\frac{1}{3} - \ln r\right) + \frac{\zeta_{2m}}{r}\right]\right\}, \quad (7.100)$$

$$\varepsilon'_{12m} = -\frac{1}{\zeta_{1m}}\left\{R_p\left[\frac{1}{r_c}\left(\frac{1}{3}-\ln r\right)-\frac{1}{r}\left(\frac{1}{3}-\ln r_c\right)\right]\left(\frac{\partial p_p}{\partial\varphi}-\frac{p_p}{\zeta_{1m}}\frac{\partial\zeta_{1m}}{\partial\varphi}\right)\right.$$
$$+\frac{p_p R_p}{r_c}\left[\frac{1}{r}-\frac{1}{r_c}\left(\frac{1}{3}-\ln r\right)\right]\frac{\partial r_c}{\partial\varphi}$$
$$\left.+\frac{u_{1c0}}{d\,\zeta_{1m}}\left[2\,c_{2m}\left(\frac{1}{3}-\ln r\right)+\frac{\zeta_{2m}}{r}\right]\frac{\partial\zeta_{1m}}{\partial\varphi}\right\}, \tag{7.101}$$

$$\varepsilon'_{13m} = -\frac{1}{\zeta_{1m}}\left\{R_p\left[\frac{1}{r_c}\left(\frac{1}{3}-\ln r\right)-\frac{1}{r}\left(\frac{1}{3}-\ln r_c\right)\right]\left(\frac{\partial p_p}{\partial\nu}-\frac{p_p}{\zeta_{1m}}\frac{\partial\zeta_{1m}}{\partial\nu}\right)\right.$$
$$+\frac{p_p R_p}{r_c}\left[\frac{1}{r}-\frac{1}{r_c}\left(\frac{1}{3}-\ln r\right)\right]\frac{\partial r_c}{\partial\nu}$$
$$\left.+\frac{u_{1c0}}{d\,\zeta_{1m}}\left[2\,c_{2m}\left(\frac{1}{3}-\ln r\right)+\frac{\zeta_{2m}}{r}\right]\frac{\partial\zeta_{1m}}{\partial\nu}\right\}, \tag{7.102}$$

$$\sigma'_{11m} =$$
$$\frac{1}{\zeta_{1m}}\left(p_p R_p\left\{\frac{1}{r_c}\left[\frac{2\,(c_{1m}+2\,c_{2m})}{3}+(c_{1m}-c_{2m})\ln r\right]-\frac{2\,c_{2m}}{r}\left(\frac{1}{3}-\ln r_c\right)\right\}\right.$$
$$\left.-\frac{2\,c_{2m}\,u_{1c0}}{d}\left[\frac{2\,(c_{1m}+2\,c_{2m})}{3}+(c_{1m}-c_{2m})\ln r+\frac{\zeta_{2m}}{r}\right]\right), \tag{7.103}$$

$$\sigma'_{22m} = \sigma'_{33m} =$$
$$-\frac{1}{\zeta_{1m}}\left(p_p R_p\left\{\frac{1}{r_c}\left[\frac{c_{1m}+2\,c_{2m}}{3}-(c_{1m}-c_{2m})\ln r\right]-\frac{c_{1m}}{r}\left(\frac{1}{3}-\ln r_c\right)\right\}\right.$$
$$\left.-\frac{u_{1c0}}{d}\left\{2\,c_{2m}\left[\frac{c_{1m}+2\,c_{2m}}{3}-(c_{1m}-c_{2m})\ln r\right]+\frac{c_{2m}\,\zeta_{2m}}{r}\right\}\right), \tag{7.104}$$

$$\sigma'_{12m} = -\frac{1}{s_{44m}\zeta_{1m}}\left\{R_p\left[\frac{1}{r_c}\left(\frac{1}{3} - \ln r\right) - \frac{1}{r}\left(\frac{1}{3} - \ln r_c\right)\right]\right.$$
$$\times\left(\frac{\partial p_p}{\partial\varphi} - \frac{p_p}{\zeta_{1m}}\frac{\partial\zeta_{1m}}{\partial\varphi}\right)$$
$$+\frac{p_p R_p}{r_c}\left[\frac{1}{r} - \frac{1}{r_c}\left(\frac{1}{3} - \ln r\right)\right]\frac{\partial r_c}{\partial\varphi}$$
$$\left.+\frac{u_{1c0}}{d\zeta_{1m}}\left[2c_{2m}\left(\frac{1}{3} - \ln r\right) + \frac{\zeta_{2m}}{r}\right]\frac{\partial\zeta_{1m}}{\partial\varphi}\right\}, \quad (7.105)$$

$$\sigma'_{13m} = -\frac{1}{s_{44m}\zeta_{1m}}\left\{R_p\left[\frac{1}{r_c}\left(\frac{1}{3} - \ln r\right) - \frac{1}{r}\left(\frac{1}{3} - \ln r_c\right)\right]\right.$$
$$\times\left(\frac{\partial p_p}{\partial\nu} - \frac{p_p}{\zeta_{1m}}\frac{\partial\zeta_{1m}}{\partial\nu}\right)$$
$$+\frac{p_p R_p}{r_c}\left[\frac{1}{r} - \frac{1}{r_c}\left(\frac{1}{3} - \ln r\right)\right]\frac{\partial r_c}{\partial\nu}$$
$$\left.+\frac{u_{1c0}}{d\zeta_{1m}}\left[2c_{2m}\left(\frac{1}{3} - \ln r\right) + \frac{\zeta_{2m}}{r}\right]\frac{\partial\zeta_{1m}}{\partial\nu}\right\}, \quad (7.106)$$

where s_{44m} is given by Equation (3.35).

7.3.2. Cell Matrix aound Void

As presented in Section 5.3, the absolute values $|u'_{1v}|$, $\left|\varepsilon'_{ijv}\right|$, $\left|\sigma'_{ijv}\right|$ $(i, j = 1,2,3)$ are required to represent increasing functions of $r \in \langle R_v, r_c\rangle$ (see Equations (3.19)–(3.23), (3.36)–(3.39)). These increasing functions exhibit maximum values on the cell surface, i.e., for $r = r_c$ (see Equations (3.15)–(3.18)). These increasing courses are ensured by the integration constants $C_{1v} \neq 0$, $C_{3v} \neq 0$, which are determined by the mandatory boundary conditions (5.5), (5.6), where $C_{2v} = 0$. With regard to Equations (5.5), (5.6), (7.4), (7.5), (7.7), (7.15), we get

$$C_{1v} = -\frac{2c_{2m}u'_{1c}}{r_c\xi_{1v}\chi_{1v}} = -\frac{2c_{2m}u_{1c0}}{d\xi_{1v}\chi_{1v}}, \quad (7.107)$$

$$C_{3v} = -\frac{u'_{1c}\chi_{2v}}{r_c\xi_{3v}\chi_{1v}} = -\frac{u_{1c0}\chi_{2v}}{d\xi_{3v}\chi_{1v}}, \tag{7.108}$$

where $u'_{1c}/r_c = u_{1c0}/d$ (see Equations (3.15)–(3.18), (5.4)); and r_c, ξ_{iv} $(i=1,3)$ are given by Equations (3.15)–(3.18), (7.8), (7.10), respectively. The displacement u_{1c0} along the axis x_j $(j=1,2,3)$ on the cell surface *I*, *II* (see Equation (5.4)), i.e., for $r = d/2$ (see Figure 2.1), is determined in Section 7.3.4. The coefficient χ_{iv} $(i=1,2)$ has the forms

$$\chi_{1v} = \frac{R_v}{r_c}\left[\frac{2(c_{1m}+2c_{2m})}{3} + (c_{1m}-c_{2m})\ln R_v\right] + 2c_{2m}\left(\frac{1}{3} - \ln r_c\right), \tag{7.109}$$

$$\chi_{2v} = R_v\left[\frac{2(c_{1m}+2c_{2m})}{3} + (c_{1m}-c_{2m})\ln R_v\right], \tag{7.110}$$

where c_{im} $(i=1,2)$ is given by Equation (3.40). With regard to Equations (7.4), (7.5), (7.11)–(7.18), (7.46), (7.47), we get

$$u'_{1v} = -\frac{r u_{1c0}}{d\chi_{1v}}\left[2c_{2m}\left(\frac{1}{3} - \ln r\right) + \frac{\chi_{2v}}{r}\right], \tag{7.111}$$

$$\varepsilon'_{11v} = \frac{2c_{2m}u_{1c0}}{d\chi_{1v}}\left(\frac{2}{3} + \ln r\right), \tag{7.112}$$

$$\varepsilon'_{22v} = \varepsilon'_{33v} = -\frac{u_{1c0}}{d\chi_{1v}}\left[2c_{2m}\left(\frac{1}{3} - \ln r\right) + \frac{\chi_{2v}}{r}\right], \tag{7.113}$$

$$\varepsilon'_{12v} = \frac{u_{1c0}}{d\chi_{1v}^2}\left[2c_{2m}\left(\frac{1}{3} - \ln r\right) + \frac{\chi_{2v}}{r}\right]\frac{\partial\chi_{1v}}{\partial\varphi}, \tag{7.114}$$

$$\varepsilon'_{13v} = \frac{u_{1c0}}{d\chi_{1v}^2}\left[2c_{2m}\left(\frac{1}{3} - \ln r\right) + \frac{\chi_{2v}}{r}\right]\frac{\partial\chi_{1v}}{\partial\nu}, \tag{7.115}$$

$$\sigma'_{11v} = \frac{2c_{2m}u_{1c0}}{d\chi_{1v}}\left[\frac{2(c_{1m}+2c_{2m})}{3} + (c_{1m}-c_{2m})\ln r + \frac{\chi_{2v}}{r}\right], \tag{7.116}$$

$$\sigma'_{22v} = \sigma'_{33v} = -\frac{u_{1c0}}{d\chi_{1v}}\left\{2c_{2m}\left[\frac{c_{1m}+2c_{2m}}{3} - (c_{1m}-c_{2m})\ln r\right] + \frac{c_{2m}\chi_{2v}}{r}\right\},$$

(7.117)

$$\sigma'_{12v} = \frac{u_{1c0}}{d\, s_{44m} \chi_{1v}^2} \left[2 c_{2m} \left(\frac{1}{3} - \ln r \right) + \frac{\chi_{2v}}{r} \right] \frac{\partial \chi_{1v}}{\partial \varphi}, \tag{7.118}$$

$$\sigma'_{13v} = \frac{u_{1c0}}{d\, s_{44m} \chi_{1v}^2} \left[2 c_{2m} \left(\frac{1}{3} - \ln r \right) + \frac{\chi_{2v}}{r} \right] \frac{\partial \chi_{1v}}{\partial \nu}, \tag{7.119}$$

where s_{44m} is given by Equation (3.35).

7.3.3. Radial Stress p_p

With regard to Equations (5.12), (6.25), (7.100), the radial stress p_p, which acts at the particle-matrix boundary, i.e., for $r = R_p$ (see Figure 2.1a), is given by Equations (7.60)–(7.63). The coefficients ζ_{3m}, ζ_{4m}, which are included in Equations (7.60), (7.61), have the forms

$$\zeta_{3m} = \frac{R_p}{\zeta_{1m}} \left[\frac{1}{r_c} \left(\frac{1}{3} - \ln R_p \right) - \frac{1}{R_p} \left(\frac{1}{3} - \ln r_c \right) \right], \tag{7.120}$$

$$\zeta_{4m} = -\frac{1}{\zeta_{1m}} \left[2 c_{2m} \left(\frac{1}{3} - \ln R_p \right) + \frac{\zeta_{2m}}{R_p} \right], \tag{7.121}$$

where r_c, c_{2m}, ζ_{1m}, ζ_{2m} are given by Equations (3.15)–(3.18), (3.40), (7.96), (7.97), respectively.

7.3.4. Displacement u_{1c0}

With regard to Equations (7.60), (7.94), (7.95), (7.107), (7.108), the integration constant C_{iq} ($i = 1,3$; $q = m,v$) has the form

$$C_{im} = \frac{1}{\xi_{im}} \left[\frac{u_{1c0} \zeta_{i1m}}{d} + \zeta_{i2m} (\beta_m - \beta_p) \right], \quad i = 1,3, \tag{7.122}$$

$$C_{iv} = \frac{u_{1c0} \zeta_{iv}}{d\, \xi_{iv}}, \quad i = 1,3, \tag{7.123}$$

where ξ_{im}, ξ_{iv} ($i = 1,3$) are given by Equations (7.8), (7.10), and β_q ($q = p,m$) is determined in Section 4.2. The coefficients ζ_{i1m}, ζ_{i2m}, ζ_{iv} ($i = 1,3$) are derived as

$$\zeta_{11m} = \frac{1}{\zeta_{1m}} \left(2c_{2m} - \frac{R_p \zeta_{pm} \zeta_{4m}}{r_c} \right), \tag{7.124}$$

$$\zeta_{31m} = \frac{1}{\zeta_{1m}} \left[R_p \zeta_{pm} \zeta_{4m} \left(\frac{1}{3} - \ln r_c \right) + \zeta_{2m} \right], \tag{7.125}$$

$$\zeta_{12m} = -\frac{R_p \zeta_{pm}}{r_c \zeta_{1m}}, \tag{7.126}$$

$$\zeta_{32m} = \frac{R_p \zeta_{pm}}{\zeta_{1m}} \left(\frac{1}{3} - \ln r_c \right), \tag{7.127}$$

$$\zeta_{1v} = -\frac{2c_{2m}}{\chi_{1v}}, \tag{7.128}$$

$$\zeta_{3v} = -\frac{\chi_{2v}}{\chi_{1v}}, \tag{7.129}$$

where r_c, ζ_{im}, χ_{iv} $(i = 1,2)$ and ζ_{pm}, ζ_{4m} are given by Equations (3.15)–(3.18), (7.96), (7.97), (7.109), (7.110) and (7.60), (7.121), respectively. The coefficients ζ_{3m}, ζ_{4m}, which are included in ζ_{pm} (see Equation (7.60)), are given by Equations (7.120), (7.121), respectively.

With regard to Equations (7.19), (7.20), (7.22), (7.25), (7.26), (7.28), (7.31), (7.122), (7.123), the elastic energy density w_q $(q = m,v)$ is given by Equations (7.72), (7.73). The coefficients $\omega_{im}^{(12)}$ $(i = 1,2,3)$, $\omega_v^{(12)}$, which are included in Equations (7.72), (7.73), are replaced by $\omega_{im}^{(13)}$ $(i = 1,2,3)$, $\omega_v^{(13)}$, respectively. The coefficients $\omega_{im}^{(13)}$ $(i = 1,2,3)$, $\omega_v^{(13)}$ are derived as

$$\begin{aligned}
\omega_{1m}^{(13)} &= \zeta_{11m}^2 u_{11m}^{(r)} + \zeta_{31m}^2 u_{33m}^{(r)} + \zeta_{11m} \zeta_{31m} u_{13m}^{(r)} \\
&+ \frac{u_{11m}^{(\varphi\nu)}}{s_{44m}} \left[\left(\frac{\partial \zeta_{11m}}{\partial \varphi} \right)^2 + \left(\frac{\partial \zeta_{11m}}{\partial \nu} \right)^2 \right] \\
&+ \frac{u_{33m}^{(\varphi\nu)}}{s_{44m}} \left[\left(\frac{\partial \zeta_{31m}}{\partial \varphi} \right)^2 + \left(\frac{\partial \zeta_{31m}}{\partial \nu} \right)^2 \right] \\
&+ \frac{u_{13m}^{(\varphi\nu)}}{s_{44m}} \left[\frac{\partial \zeta_{11m}}{\partial \varphi} \frac{\partial \zeta_{31m}}{\partial \varphi} + \frac{\partial \zeta_{11m}}{\partial \nu} \frac{\partial \zeta_{31m}}{\partial \nu} \right],
\end{aligned} \tag{7.130}$$

$$\omega_{2m}^{(13)} = 2\left[\zeta_{11m}\zeta_{12m}u_{11m}^{(r)} + \zeta_{31m}\zeta_{32m}u_{33m}^{(r)}\right] + \left(\zeta_{11m}\zeta_{32m} + \zeta_{12m}\zeta_{31m}\right)u_{13m}^{(r)}$$
$$+ \frac{2u_{11m}^{(\varphi\nu)}}{s_{44m}}\left(\frac{\partial\zeta_{11m}}{\partial\varphi}\frac{\partial\zeta_{12m}}{\partial\varphi} + \frac{\partial\zeta_{11m}}{\partial\nu}\frac{\partial\zeta_{12m}}{\partial\nu}\right)$$
$$+ \frac{2u_{33m}^{(\varphi\nu)}}{s_{44m}}\left(\frac{\partial\zeta_{31m}}{\partial\varphi}\frac{\partial\zeta_{32m}}{\partial\varphi} + \frac{\partial\zeta_{31m}}{\partial\nu}\frac{\partial\zeta_{32m}}{\partial\nu}\right)$$
$$+ \frac{u_{13m}^{(\varphi\nu)}}{s_{44m}}\left(\frac{\partial\zeta_{11m}}{\partial\varphi}\frac{\partial\zeta_{32m}}{\partial\varphi} + \frac{\partial\zeta_{12m}}{\partial\varphi}\frac{\partial\zeta_{31m}}{\partial\varphi}\right.$$
$$\left. + \frac{\partial\zeta_{11m}}{\partial\nu}\frac{\partial\zeta_{32m}}{\partial\nu} + \frac{\partial\zeta_{12m}}{\partial\nu}\frac{\partial\zeta_{31m}}{\partial\nu}\right), \tag{7.131}$$

$$\omega_{3m}^{(13)} = \zeta_{12m}^{2}u_{11m}^{(r)} + \zeta_{32m}^{2}u_{33m}^{(r)} + \zeta_{12m}\zeta_{32m}u_{13m}^{(r)}$$
$$+ \frac{u_{11m}^{(\varphi\nu)}}{s_{44m}}\left[\left(\frac{\partial\zeta_{12m}}{\partial\varphi}\right)^2 + \left(\frac{\partial\zeta_{12m}}{\partial\nu}\right)^2\right]$$
$$+ \frac{u_{33m}^{(\varphi\nu)}}{s_{44m}}\left[\left(\frac{\partial\zeta_{32m}}{\partial\varphi}\right)^2 + \left(\frac{\partial\zeta_{32m}}{\partial\nu}\right)^2\right]$$
$$+ \frac{u_{13m}^{(\varphi\nu)}}{s_{44m}}\left[\frac{\partial\zeta_{12m}}{\partial\varphi}\frac{\partial\zeta_{32m}}{\partial\varphi} + \frac{\partial\zeta_{12m}}{\partial\nu}\frac{\partial\zeta_{32m}}{\partial\nu}\right], \tag{7.132}$$

$$\omega_{v}^{(13)} = \zeta_{1v}^{2}u_{11m}^{(r)} + \zeta_{3v}^{2}u_{33m}^{(r)} + \zeta_{1v}\zeta_{3v}u_{13m}^{(r)} + \frac{u_{11m}^{(\varphi\nu)}}{s_{44m}}\left[\left(\frac{\partial\zeta_{1v}}{\partial\varphi}\right)^2 + \left(\frac{\partial\zeta_{1v}}{\partial\nu}\right)^2\right]$$
$$+ \frac{u_{33m}^{(\varphi\nu)}}{s_{44m}}\left[\left(\frac{\partial\zeta_{3v}}{\partial\varphi}\right)^2 + \left(\frac{\partial\zeta_{3v}}{\partial\nu}\right)^2\right] + \frac{u_{13m}^{(\varphi\nu)}}{s_{44m}}\left[\frac{\partial\zeta_{1v}}{\partial\varphi}\frac{\partial\zeta_{3v}}{\partial\varphi} + \frac{\partial\zeta_{1v}}{\partial\nu}\frac{\partial\zeta_{3v}}{\partial\nu}\right], \tag{7.133}$$

where s_{44m} is given by Equation (3.35).

With regard to Equations (3.47), (7.72), the elastic energy W_m, which is accumulated in the cell matrix I, is given by Equation (7.78). With regard to Equations (7.72), (7.130)–(7.132), the coefficient $\Omega_{im}^{(12)}$ $(i = 1,2,3)$, which is included in Equation (7.78), is replaced by $\Omega_{im}^{(13)}$ $(i = 1,2,3)$. The coefficient $\Omega_{im}^{(13)}$ $(i = 1,2,3)$ has the form

$$\Omega_{1m}^{(13)} = \int_0^{\pi/2}\int_0^{\pi/2}\int_{R_p}^{r_c} \omega_{1m}^{(13)} r^2\, dr\, d\varphi\, d\nu$$
$$= \int_0^{\pi/2}\int_0^{\pi/2} \Bigg\{ \zeta_{11m}^2 U_{11m}^{(r)} + \zeta_{31m}^2 U_{33m}^{(r)} + \zeta_{11m}\zeta_{31m} U_{13m}^{(r)}$$
$$+ \frac{U_{11m}^{(\varphi\nu)}}{s_{44m}} \left[\left(\frac{\partial \zeta_{11m}}{\partial \varphi}\right)^2 + \left(\frac{\partial \zeta_{11m}}{\partial \nu}\right)^2 \right]$$
$$+ \frac{U_{33m}^{(\varphi\nu)}}{s_{44m}} \left[\left(\frac{\partial \zeta_{31m}}{\partial \varphi}\right)^2 + \left(\frac{\partial \zeta_{31m}}{\partial \nu}\right)^2 \right]$$
$$+ \frac{U_{13m}^{(\varphi\nu)}}{s_{44m}} \left[\frac{\partial \zeta_{11m}}{\partial \varphi}\frac{\partial \zeta_{31m}}{\partial \varphi} + \frac{\partial \zeta_{11m}}{\partial \nu}\frac{\partial \zeta_{31m}}{\partial \nu} \right] \Bigg\}\, d\varphi\, d\nu, \quad (7.134)$$

$$\Omega_{2m}^{(13)} = \int_0^{\pi/2}\int_0^{\pi/2}\int_{R_p}^{r_c} \omega_{2m}^{(13)} r^2\, dr\, d\varphi\, d\nu$$
$$= \int_0^{\pi/2}\int_0^{\pi/2} \Bigg\{ 2\left[\zeta_{11m}\zeta_{12m} U_{11m}^{(r)} + \zeta_{31m}\zeta_{32m} U_{33m}^{(r)}\right]$$
$$+ \left(\zeta_{11m}\zeta_{32m} + \zeta_{12m}\zeta_{31m}\right) U_{13m}^{(r)}$$
$$+ \frac{2U_{11m}^{(\varphi\nu)}}{s_{44m}} \left(\frac{\partial \zeta_{11m}}{\partial \varphi}\frac{\partial \zeta_{12m}}{\partial \varphi} + \frac{\partial \zeta_{11m}}{\partial \nu}\frac{\partial \zeta_{12m}}{\partial \nu} \right)$$
$$+ \frac{2U_{33m}^{(\varphi\nu)}}{s_{44m}} \left(\frac{\partial \zeta_{31m}}{\partial \varphi}\frac{\partial \zeta_{32m}}{\partial \varphi} + \frac{\partial \zeta_{31m}}{\partial \nu}\frac{\partial \zeta_{32m}}{\partial \nu} \right)$$
$$+ \frac{U_{13m}^{(\varphi\nu)}}{s_{44m}} \left(\frac{\partial \zeta_{11m}}{\partial \varphi}\frac{\partial \zeta_{32m}}{\partial \varphi} + \frac{\partial \zeta_{12m}}{\partial \varphi}\frac{\partial \zeta_{31m}}{\partial \varphi} \right)$$
$$+ \frac{U_{13m}^{(\varphi\nu)}}{s_{44m}} \left(\frac{\partial \zeta_{11m}}{\partial \nu}\frac{\partial \zeta_{32m}}{\partial \nu} + \frac{\partial \zeta_{12m}}{\partial \nu}\frac{\partial \zeta_{31m}}{\partial \nu} \right) \Bigg\}\, d\varphi\, d\nu, \quad (7.135)$$

$$\Omega_{3m}^{(13)} = \int_0^{\pi/2}\int_0^{\pi/2}\int_{R_p}^{r_c} \omega_{3m}^{(13)} r^2 \, dr \, d\varphi \, d\nu$$

$$= \int_0^{\pi/2}\int_0^{\pi/2} \left\{ \zeta_{12m}^2 U_{11m}^{(r)} + \zeta_{32m}^2 U_{33m}^{(r)} + \zeta_{12m}\zeta_{32m} U_{13m}^{(r)} + \frac{U_{11m}^{(\varphi\nu)}}{s_{44m}} \left[\left(\frac{\partial \zeta_{12m}}{\partial \varphi}\right)^2 + \left(\frac{\partial \zeta_{12m}}{\partial \nu}\right)^2 \right] + \frac{U_{33m}^{(\varphi\nu)}}{s_{44m}} \left[\left(\frac{\partial \zeta_{32m}}{\partial \varphi}\right)^2 + \left(\frac{\partial \zeta_{32m}}{\partial \nu}\right)^2 \right] + \frac{U_{13m}^{(\varphi\nu)}}{s_{44m}} \left[\frac{\partial \zeta_{12m}}{\partial \varphi}\frac{\partial \zeta_{32m}}{\partial \varphi} + \frac{\partial \zeta_{12m}}{\partial \nu}\frac{\partial \zeta_{32m}}{\partial \nu} \right] \right\} d\varphi \, d\nu, \quad (7.136)$$

where $U_{ijm}^{(x)}$ $(i, j = 1,3;\ x = r, \varphi\nu)$, $U_{11m}^{(r)}$, $U_{11m}^{(\varphi\nu)}$ are given by Equations (7.82), (7.83), (7.86), respectively. With regard to Equations (7.22), (7.25), (7.28), (7.31), (7.82), we get

$$U_{13m}^{(r)} = c_{1m}\left(r_c^2 - R_p^2\right), \quad (7.137)$$

$$U_{33m}^{(r)} = c_{1m}\left(r_c - R_p\right), \quad (7.138)$$

$$U_{13m}^{(\varphi\nu)} = r_c^2\left(\frac{5}{6} - \ln r_c\right) - R_p^2\left(\frac{5}{6} - \ln R_p\right), \quad (7.139)$$

$$U_{33m}^{(\varphi\nu)} = r_c - R_p, \quad (7.140)$$ [1]

where r_c, c_{im} $(i = 1,3)$ are given by Equations (3.15)–(3.18), (3.40), respectively.

With regard to Equations (3.48), (7.73), the elastic energy W_v, which is accumulated in the cell matrix *II*, is given by Equation (7.89). With regard to Equations (7.73), (7.133), the coefficient $\Omega_v^{(12)}$, which is included in Equation (7.89), is replaced by $\Omega_v^{(13)}$. The coefficient $\Omega_v^{(13)}$ has the form

[1] Môj milovaný otecko zomrel. Môj najdrahší otecko, odpočívaj v pokoji v nebi.
My beloved daddy died. My dearest daddy, rest in peace in heaven.

$$\Omega_{v}^{(13)} = \int_{0}^{\pi/2}\int_{0}^{\pi/2}\int_{R_p}^{r_c} \omega_{v}^{(13)} r^2 \, dr \, d\varphi \, d\nu$$

$$= \int_{0}^{\pi/2}\int_{0}^{\pi/2} \left\{ \zeta_{1v}^2 U_{11v}^{(r)} + \zeta_{3v}^2 U_{33v}^{(r)} + \zeta_{1v}\zeta_{3v} U_{13v}^{(r)} \right.$$
$$+ \frac{U_{11v}^{(\varphi\nu)}}{s_{44m}} \left[\left(\frac{\partial \zeta_{1v}}{\partial \varphi} \right)^2 + \left(\frac{\partial \zeta_{1v}}{\partial \nu} \right)^2 \right]$$
$$+ \frac{U_{33v}^{(\varphi\nu)}}{s_{44m}} \left[\left(\frac{\partial \zeta_{3v}}{\partial \varphi} \right)^2 + \left(\frac{\partial \zeta_{3v}}{\partial \nu} \right)^2 \right]$$
$$\left. + \frac{U_{13v}^{(\varphi\nu)}}{s_{44m}} \left[\frac{\partial \zeta_{1v}}{\partial \varphi} \frac{\partial \zeta_{3v}}{\partial \varphi} + \frac{\partial \zeta_{1v}}{\partial \nu} \frac{\partial \zeta_{3v}}{\partial \nu} \right] \right\} d\varphi \, d\nu. \qquad (7.141)$$

With regard to Equations (7.82), (7.91), $U_{ijv}^{(x)}$ $(i, j = 1,3;\ x = r,\ \varphi\nu)$ is given by Equations (7.137)–(7.140). In case of $U_{ijv}^{(x)}$, the radius R_p, which is included in Equations (7.137)–(7.140), is replaced by R_v.

With regard to Equations (5.13), (6.81), (7.78), (7.89), the displacement u_{1c0} along the axis x_i $(i = 1,2,3)$ on the cubic cell surface *I, II*, i.e., for $r = d/2$ (see Figure 2.1a), is given by Equations (6.89), (7.92), (7.93).

The coefficients $\Omega_{1m}^{(12)}$, $\Omega_{2m}^{(12)}$, $\Omega_{v}^{(12)}$, which are included in Equations (7.92), (7.93), are replaced by $\Omega_{1m}^{(13)}$, $\Omega_{2m}^{(13)}$, $\Omega_{v}^{(13)}$ (see Equations (7.134), (7.135), (7.141)), respectively.

7.4. Integration Constants $C_{2q} \neq 0, C_{3q} \neq 0, C_{1q} = 0$

7.4.1. Cell Matrix Around Particle

As presented above (see page 55), $u_{2rm} = u_{2rm}(r)$ is a decreasing function of the variable r due to $c_{3m} < 0$, and u_{3rm} is a constant function with respect to the variable r (see Equations (3.40), (7.6), (7.7)).

As presented in Section 5.2, the absolute values $|u'_{1m}|$, $\left|\varepsilon'_{ijm}\right|$, $\left|\sigma'_{ijm}\right|$ $(i, j = 1,2,3)$ are required to represent decreasing functions of $r \in \langle R_p, r_c \rangle$ (see

Equations (3.19)–(3.23), (3.36)–(3.39)). These decreasing functions exhibit maximum values on the particle-matrix boundary, i.e., for $r = R_p$. These decreasing courses are ensured by the integration constants $C_{2m} \neq 0$, $C_{3m} \neq 0$, which are determined by the mandatory boundary conditions (5.3), (5.4), where $C_{1m} = 0$. With regard to Equations (5.3), (5.4), (7.4), (7.6), (7.7), (7.15), we get

$$C_{2m} = -\frac{1}{\xi_{2m}\zeta_{1m}r_c^{c_{3m}-1}}\left(\frac{p_p R_p}{r_c} - \frac{2c_{2m}u'_{1c}}{r_c}\right) = -\frac{1}{\xi_{2m}\zeta_{1m}r_c^{c_{3m}-1}}\left(\frac{p_p R_p}{r_c} - \frac{2c_{2m}u_{1c0}}{d}\right), \tag{7.142}$$

$$C_{3m} = \frac{R_p}{\xi_{3m}\zeta_{1m}}\left(p_p + \frac{u'_{1c}\zeta_{2m}}{r_c}\right) = \frac{R_p}{\xi_{3m}\zeta_{1m}}\left(p_p + \frac{u_{1c0}\zeta_{2m}}{d}\right), \tag{7.143}$$

where $u'_{1c}/r_c = u_{1c0}/d$ (see Equations (3.15)–(3.18), (5.4)); and r_c, ξ_{im} $(i = 2,3)$ are given by Equations (3.15)–(3.18), (7.9), (7.10), respectively. The radial stress p_p, which acts at the particle-matrix boundary, i.e., for $r = R_p$ (see Figure 2.1a), is determined in Section 7.4.3. The displacement u_{1c0} along the axis x_j $(j = 1,2,3)$ on the cell surface *I*, *II* (see Equation (5.4)), i.e., for $r = d/2$ (see Figure 2.1), is determined in Section 7.4.4. The coefficient ζ_{im} $(i = 1,2)$ has the forms

$$\zeta_{1m} = [c_{3m}(c_{1m} + c_{2m}) - 2c_{2m}]\left(\frac{R_p}{r_c}\right)^{c_{3m}} + 2c_{2m}, \tag{7.144}$$

$$\zeta_{2m} = [c_{3m}(c_{1m} + c_{2m}) - 2c_{2m}]\left(\frac{R_p}{r_c}\right)^{c_{3m}-1}, \tag{7.145}$$

where c_{im} $(i = 1,2)$ is given by Equation (3.40). With regard to Equations (7.4), (7.6), (7.7), (7.11)–(7.18), (7.142), (7.143), we get

$$u'_{1m} = \frac{1}{\zeta_{1m}}\left\{p_p R_p\left[1 - \left(\frac{r}{r_c}\right)^{c_{3m}}\right] + \frac{u_{1c0}}{d}\left[R_p\zeta_{2m} + 2c_{2m}r\left(\frac{r}{r_c}\right)^{c_{3m}-1}\right]\right\}, \tag{7.146}$$

$$\varepsilon'_{11m} = -\frac{c_{3m}}{\zeta_{1m}}\left(\frac{p_p R_p}{r_c} - \frac{2c_{2m}u_{1c0}}{d}\right)\left(\frac{r}{r_c}\right)^{c_{3m}-1}, \tag{7.147}$$

$$\varepsilon'_{22m} = \varepsilon'_{33m} = \frac{1}{r\zeta_{1m}} \left\{ p_p R_p \left[1 - \left(\frac{r}{r_c} \right)^{c_{3m}} \right] \right.$$
$$\left. + \frac{u_{1c0}}{d} \left[R_p \zeta_{2m} + 2 c_{2m} r \left(\frac{r}{r_c} \right)^{c_{3m}-1} \right] \right\}, \qquad (7.148)$$

$$\varepsilon'_{12m} = \frac{1}{\zeta_{1m}} \left(R_p \left\{ \left[\frac{1}{r} - \frac{1}{r_c} \left(\frac{r}{r_c} \right)^{c_{3m}-1} \right] \left(\frac{\partial p_p}{\partial \varphi} - \frac{p_p}{\zeta_{1m}} \frac{\partial \zeta_{1m}}{\partial \varphi} \right) \right. \right.$$
$$\left. + \frac{p_p c_{3m}}{r_c^2} \left(\frac{r}{r_c} \right)^{c_{3m}-1} \frac{\partial r_c}{\partial \varphi} \right\}$$
$$+ \frac{u_{1c0}}{d} \left[\frac{R_p}{r} \left(\frac{\partial \zeta_{2m}}{\partial \varphi} - \frac{\zeta_{2m}}{\zeta_{1m}} \frac{\partial \zeta_{1m}}{\partial \varphi} \right) \right.$$
$$\left. \left. + 2 c_{2m} \left(\frac{1 - c_{3m}}{r_c} \frac{\partial r_c}{\partial \varphi} - \frac{1}{\zeta_{1m}} \frac{\partial \zeta_{1m}}{\partial \varphi} \right) \left(\frac{r}{r_c} \right)^{c_{3m}-1} \right] \right), \qquad (7.149)$$

$$\varepsilon'_{13m} = \frac{1}{\zeta_{1m}} \left(R_p \left\{ \left[\frac{1}{r} - \frac{1}{r_c} \left(\frac{r}{r_c} \right)^{c_{3m}-1} \right] \left(\frac{\partial p_p}{\partial \nu} - \frac{p_p}{\zeta_{1m}} \frac{\partial \zeta_{1m}}{\partial \nu} \right) \right. \right.$$
$$\left. + \frac{p_p c_{3m}}{r_c^2} \left(\frac{r}{r_c} \right)^{c_{3m}-1} \frac{\partial r_c}{\partial \nu} \right\}$$
$$+ \frac{u_{1c0}}{d} \left[\frac{R_p}{r} \left(\frac{\partial \zeta_{2m}}{\partial \nu} - \frac{\zeta_{2m}}{\zeta_{1m}} \frac{\partial \zeta_{1m}}{\partial \nu} \right) \right.$$
$$\left. \left. + 2 c_{2m} \left(\frac{1 - c_{3m}}{r_c} \frac{\partial r_c}{\partial \nu} - \frac{1}{\zeta_{1m}} \frac{\partial \zeta_{1m}}{\partial \nu} \right) \left(\frac{r}{r_c} \right)^{c_{3m}-1} \right] \right), \qquad (7.150)$$

$$\sigma'_{11m} =$$
$$- \frac{1}{\zeta_{1m}} \left(p_p R_p \left\{ \frac{1}{r_c} \left[c_{3m} (c_{1m} + c_{2m}) - 2 c_{2m} \right] \left(\frac{r}{r_c} \right)^{c_{3m}-1} + \frac{2 c_{2m}}{r} \right\} \right.$$

$$- \frac{2c_{2m}u_{1c0}}{d}\left\{\left[c_{3m}\left(c_{1m}+c_{2m}\right)-2c_{2m}\right]\left(\frac{r}{r_c}\right)^{c_{3m}-1} - \frac{R_p\zeta_{2m}}{r}\right\}\Bigg), \tag{7.151}$$

$$\sigma'_{22m} = \sigma'_{33m} = - \frac{1}{\zeta_{1m}}\left\{p_p R_p\left[\frac{c_{1m}-c_{2m}c_{3m}}{r_c}\left(\frac{r}{r_c}\right)^{c_{3m}-1} - \frac{c_{1m}}{r}\right] - \frac{u_{1c0}}{d}\left[2c_{2m}\left(c_{1m}-c_{2m}c_{3m}\right)\left(\frac{r}{r_c}\right)^{c_{3m}-1} + \frac{R_p c_{1m}\zeta_{2m}}{r}\right]\right\}, \tag{7.152}$$

$$\sigma'_{12m} = \frac{1}{s_{44m}\zeta_{1m}}\left(R_p\left\{\left[\frac{1}{r}-\frac{1}{r_c}\left(\frac{r}{r_c}\right)^{c_{3m}-1}\right]\left(\frac{\partial p_p}{\partial\varphi}-\frac{p_p}{\zeta_{1m}}\frac{\partial\zeta_{1m}}{\partial\varphi}\right) + \frac{p_p c_{3m}}{r_c^2}\left(\frac{r}{r_c}\right)^{c_{3m}-1}\frac{\partial r_c}{\partial\varphi}\right\} + \frac{u_{1c0}}{d}\left[\frac{R_p}{r}\left(\frac{\partial\zeta_{2m}}{\partial\varphi}-\frac{\zeta_{2m}}{\zeta_{1m}}\frac{\partial\zeta_{1m}}{\partial\varphi}\right) + 2c_{2m}\left(\frac{1-c_{3m}}{r_c}\frac{\partial r_c}{\partial\varphi}-\frac{1}{\zeta_{1m}}\frac{\partial\zeta_{1m}}{\partial\varphi}\right)\left(\frac{r}{r_c}\right)^{c_{3m}-1}\right]\right), \tag{7.153}$$

$$\sigma'_{13m} = \frac{1}{s_{44m}\zeta_{1m}}\left(R_p\left\{\left[\frac{1}{r}-\frac{1}{r_c}\left(\frac{r}{r_c}\right)^{c_{3m}-1}\right]\left(\frac{\partial p_p}{\partial\nu}-\frac{p_p}{\zeta_{1m}}\frac{\partial\zeta_{1m}}{\partial\nu}\right)\right.\right.$$

$$+ \frac{p_p c_{3m}}{r_c^2} \left(\frac{r}{r_c} \right)^{c_{3m}-1} \frac{\partial r_c}{\partial \nu} \Bigg\}$$

$$+ \frac{u_{1c0}}{d} \left[\frac{R_p}{r} \left(\frac{\partial \zeta_{2m}}{\partial \nu} - \frac{\zeta_{2m}}{\zeta_{1m}} \frac{\partial \zeta_{1m}}{\partial \nu} \right) \right.$$

$$\left. \left. + 2 c_{2m} \left(\frac{1 - c_{3m}}{r_c} \frac{\partial r_c}{\partial \nu} - \frac{1}{\zeta_{1m}} \frac{\partial \zeta_{1m}}{\partial \nu} \right) \left(\frac{r}{r_c} \right)^{c_{3m}-1} \right] \right), \tag{7.154}$$

where s_{44m} is given by Equation (3.35).

7.4.2. Cell Matrix around Void

As presented in Section 5.3, the absolute values $|u'_{1v}|$, $\left|\varepsilon'_{ijv}\right|$, $\left|\sigma'_{ijv}\right|$ $(i, j = 1,2,3)$ are required to represent increasing functions of $r \in \langle R_v, r_c \rangle$ (see Equations (3.19)–(3.23), (3.36)–(3.39)). These increasing functions exhibit maximum values on the cell surface, i.e., for $r = r_c$ (see Equations (3.15)–(3.18)). These increasing courses are ensured by the integration constants $C_{2v} \neq 0$, $C_{3v} \neq 0$, which are determined by the mandatory boundary conditions (5.5), (5.6), where $C_{1v} = 0$. With regard to Equations (5.5), (5.6), (7.4), (7.6), (7.7), (7.15), we get

$$C_{2v} = -\frac{2 c_{2m} u'_{1c}}{\xi_{2v} \chi_{1v} r_c^{c_{3m}}} = -\frac{2 c_{2m} u_{1c0}}{d \xi_{2v} \chi_{1v} r_c^{c_{3m}-1}}, \tag{7.155}$$

$$C_{3v} = \frac{R_v u'_{1c} \chi_{2v}}{\xi_{3v} \chi_{1v} r_c} = \frac{R_v u_{1c0} \chi_{2v}}{d \xi_{3v} \chi_{1v}}, \tag{7.156}$$

where $u'_{1c}/r_c = u_{1c0}/d$ (see Equations (3.15)–(3.18), (5.4)); and r_c, ξ_{iv} $(i = 2,3)$ are given by Equations (3.15)–(3.18), (7.9), (7.10), respectively. The displacement u_{1c0} along the axis x_j $(j = 1,2,3)$ on the cell surface *I*, *II* (see Equation (5.4)), i.e., for $r = d/2$ (see Figure 2.1), is determined in Section 7.4.4. The coefficient χ_{iv} $(i = 1,2)$ has the forms

$$\chi_{1v} = [c_{3m}(c_{1m} + c_{2m}) - 2 c_{2m}] \left(\frac{R_v}{r_c} \right)^{c_{3m}} + 2 c_{2m}, \tag{7.157}$$

$$\chi_{2v} = [c_{3m}(c_{1m} + c_{2m}) - 2 c_{2m}] \left(\frac{R_v}{r_c} \right)^{c_{3m}-1}, \tag{7.158}$$

where c_{im} $(i = 1,2)$ is given by Equation (3.40). With regard to Equations (7.4), (7.6), (7.7), (7.11)–(7.18), (7.158), (7.159), we get

$$u'_{1v} = -\frac{u_{1c0}}{d\chi_{1v}}\left[R_v\chi_{2v} + 2c_{2m}r\left(\frac{r}{r_c}\right)^{c_{3m}-1}\right], \tag{7.159}$$

$$\varepsilon'_{11v} = -\frac{2c_{2m}c_{3m}u_{1c0}}{d\chi_{1v}}\left(\frac{r}{r_c}\right)^{c_{3m}-1}, \tag{7.160}$$

$$\varepsilon'_{22v} = \varepsilon'_{33v} = -\frac{u_{1c0}}{d\chi_{1v}}\left[\frac{R_v\chi_{2v}}{r} + 2c_{2m}\left(\frac{r}{r_c}\right)^{c_{3m}-1}\right], \tag{7.161}$$

$$\varepsilon'_{12v} = -\frac{u_{1c0}}{d\chi_{1v}}\left[\frac{R_v}{r}\left(\frac{\partial\chi_{2v}}{\partial\varphi} - \frac{\chi_{2v}}{\chi_{1v}}\frac{\partial\chi_{1v}}{\partial\varphi}\right)\right.$$
$$\left. + 2c_{2m}\left(\frac{1-c_{3m}}{r_c}\frac{\partial r_c}{\partial\varphi} - \frac{1}{\chi_{1v}}\frac{\partial\chi_{1v}}{\partial\varphi}\right)\left(\frac{r}{r_c}\right)^{c_{3m}-1}\right], \tag{7.162}$$

$$\varepsilon'_{12v} = -\frac{u_{1c0}}{d\chi_{1v}}\left[\frac{R_v}{r}\left(\frac{\partial\chi_{2v}}{\partial\nu} - \frac{\chi_{2v}}{\chi_{1v}}\frac{\partial\chi_{1v}}{\partial\nu}\right)\right.$$
$$\left. + 2c_{2m}\left(\frac{1-c_{3m}}{r_c}\frac{\partial r_c}{\partial\nu} - \frac{1}{\chi_{1v}}\frac{\partial\chi_{1v}}{\partial\nu}\right)\left(\frac{r}{r_c}\right)^{c_{3m}-1}\right], \tag{7.163}$$

$$\sigma'_{11v} = -\frac{2c_{2m}u_{1c0}}{d\chi_{1v}}\left\{[c_{3m}(c_{1m} + c_{2m}) - 2c_{2m}]\left(\frac{r}{r_c}\right)^{c_{3m}-1} - \frac{R_v\chi_{2v}}{r}\right\}, \tag{7.164}$$

$$\sigma'_{22v} = \sigma'_{33v} = -\frac{u_{1c0}}{d\chi_{1v}}\left[2c_{2m}(c_{1m} - c_{2m}c_{3m})\left(\frac{r}{r_c}\right)^{c_{3m}-1} + \frac{R_v c_{1m}\chi_{2v}}{r}\right], \tag{7.165}$$

$$\sigma'_{12v} = -\frac{u_{1c0}}{d\,s_{44m}\chi_{1v}}\left[\frac{R_v}{r}\left(\frac{\partial\chi_{2v}}{\partial\varphi} - \frac{\chi_{2v}}{\chi_{1v}}\frac{\partial\chi_{1v}}{\partial\varphi}\right)\right.$$

$$+ 2c_{2m}\left(\frac{1-c_{3m}}{r_c}\frac{\partial r_c}{\partial \varphi} - \frac{1}{\chi_{1v}}\frac{\partial \chi_{1v}}{\partial \varphi}\right)\left(\frac{r}{r_c}\right)^{c_{3m}-1}\Bigg], \tag{7.166}$$

$$\sigma'_{13v} = -\frac{u_{1c0}}{d\, s_{44m}\chi_{1v}}\Bigg[\frac{R_v}{r}\left(\frac{\partial \chi_{2v}}{\partial \nu} - \frac{\chi_{2v}}{\chi_{1v}}\frac{\partial \chi_{1v}}{\partial \nu}\right) + 2c_{2m}\left(\frac{1-c_{3m}}{r_c}\frac{\partial r_c}{\partial \nu} - \frac{1}{\chi_{1v}}\frac{\partial \chi_{1v}}{\partial \nu}\right)\left(\frac{r}{r_c}\right)^{c_{3m}-1}\Bigg], \tag{7.167}$$

where s_{44m} is given by Equation (3.35).

7.4.3. Radial Stress p_p

With regard to Equations (5.12), (6.25), (7.148), the radial stress p_p, which acts at the particle-matrix boundary, i.e., for $r = R_p$ (see Figure 2.1a), is given by Equations (7.60)–(7.63). The coefficients ζ_{3m}, ζ_{4m}, which are included in Equations (7.60), (7.61), have the forms

$$\zeta_{3m} = -\frac{1}{\zeta_{1m}}\left[1-\left(\frac{R_p}{r_c}\right)^{c_{3m}}\right], \tag{7.168}$$

$$\zeta_{4m} = -\frac{1}{\zeta_{1m}}\left[\zeta_{2m} + 2c_{2m}\left(\frac{R_p}{r_c}\right)^{c_{3m}-1}\right], \tag{7.169}$$

where r_c, c_{im} $(i-2,3)$, ζ_{1m}, ζ_{2m} are given by Equations (3.15)–(3.18), (3.40), (7.144), (7.145), respectively.

7.4.4. Displacement u_{1c0}

With regard to Equations (7.60), (7.142), (7.143), (7.155), (7.156), the integration constant C_{iq} $(i = 2,3;\ q = m,v)$ has the form

$$C_{im} = \frac{1}{\xi_{im}}\left[\frac{u_{1c0}\,\zeta_{i1m}}{d} + \zeta_{i2m}(\beta_m - \beta_p)\right], \quad i = 2,3, \tag{7.170}$$

$$C_{iv} = \frac{u_{1c0}\,\zeta_{iv}}{d\,\xi_{iv}}, \quad i = 2,3, \tag{7.171}$$

where ξ_{im}, ξ_{iv} $(i = 2{,}3)$ are given by Equations (7.9), (7.10), and β_q $(q = p{,}m)$ is determined in Section 4.2. The coefficients ζ_{i1m}, ζ_{i2m}, ζ_{iv} $(i = 2{,}3)$ are derived as

$$\zeta_{21m} = \frac{1}{\zeta_{1m}\,r_c^{c_{3m}-1}}\left(2\,c_{2m} - \frac{R_p\,\zeta_{pm}\,\zeta_{4m}}{r_c}\right), \tag{7.172}$$

$$\zeta_{31m} = \frac{R_p\,(\zeta_{pm}\,\zeta_{4m} + \zeta_{2m})}{\zeta_{1m}}, \tag{7.173}$$

$$\zeta_{22m} = -\,\frac{R_p\,\zeta_{pm}}{\zeta_{1m}\,r_c^{c_{3m}}}, \tag{7.174}$$

$$\zeta_{32m} = \frac{R_p\,\zeta_{pm}}{\zeta_{1m}}, \tag{7.175}$$

$$\zeta_{2v} = -\,\frac{2\,c_{2m}}{\chi_{1v}\,r_c^{c_{3m}-1}}, \tag{7.176}$$

$$\zeta_{3v} = \frac{R_v\,\chi_{2v}}{\chi_{1v}}, \tag{7.177}$$

where r_c, c_{3m}, ζ_{im}, χ_{iv} $(i = 1{,}2)$ and ζ_{pm}, ζ_{4m} are given by Equations (3.15)–(3.18), (3.40), (7.144), (7.145), (7.157), (7.158) and (7.60), (7.169), respectively. The coefficients ζ_{3m}, ζ_{4m}, which are included in ζ_{pm} (see Equation (7.60)), are given by Equations (7.168), (7.169), respectively.

With regard to Equations (7.19), (7.23)–(7.25), (7.29)–(7.31), (7.170), (7.171), the elastic energy density w_q $(q = m{,}v)$ is given by Equations (7.72), (7.73). The coefficients $\omega_{im}^{(12)}$ $(i = 1{,}2{,}3)$, $\omega_v^{(12)}$, which are included in Equations (7.72), (7.73), are replaced by $\omega_{im}^{(23)}$ $(i = 1{,}2{,}3)$, $\omega_v^{(23)}$, respectively. The coefficients $\omega_{im}^{(23)}$ $(i = 1{,}2{,}3)$, $\omega_v^{(23)}$ are derived as

$$\begin{aligned}\omega_{1m}^{(23)} &= \zeta_{21m}^2\,u_{22m}^{(r)} + \zeta_{31m}^2\,u_{33m}^{(r)} \\ &+ \frac{u_{22m}^{(\varphi\nu)}}{s_{44m}}\left[\left(\frac{\partial\zeta_{21m}}{\partial\varphi}\right)^2 + \left(\frac{\partial\zeta_{21m}}{\partial\nu}\right)^2\right]\end{aligned}$$

$$+\frac{u_{33m}^{(\varphi\nu)}}{s_{44m}}\left[\left(\frac{\partial\zeta_{31m}}{\partial\varphi}\right)^2+\left(\frac{\partial\zeta_{31m}}{\partial\nu}\right)^2\right]$$

$$+\frac{u_{23m}^{(\varphi\nu)}}{s_{44m}}\left[\frac{\partial\zeta_{21m}}{\partial\varphi}\frac{\partial\zeta_{31m}}{\partial\varphi}+\frac{\partial\zeta_{21m}}{\partial\nu}\frac{\partial\zeta_{31m}}{\partial\nu}\right], \qquad (7.178)$$

$$\omega_{2m}^{(23)}=2\left[\zeta_{21m}\zeta_{22m}u_{22m}^{(r)}+\zeta_{31m}\zeta_{32m}u_{33m}^{(r)}\right]$$

$$+\frac{2u_{22m}^{(\varphi\nu)}}{s_{44m}}\left(\frac{\partial\zeta_{21m}}{\partial\varphi}\frac{\partial\zeta_{22m}}{\partial\varphi}+\frac{\partial\zeta_{21m}}{\partial\nu}\frac{\partial\zeta_{22m}}{\partial\nu}\right)$$

$$+\frac{2u_{33m}^{(\varphi\nu)}}{s_{44m}}\left(\frac{\partial\zeta_{31m}}{\partial\varphi}\frac{\partial\zeta_{32m}}{\partial\varphi}+\frac{\partial\zeta_{31m}}{\partial\nu}\frac{\partial\zeta_{32m}}{\partial\nu}\right)$$

$$+\frac{u_{23m}^{(\varphi\nu)}}{s_{44m}}\left(\frac{\partial\zeta_{21m}}{\partial\varphi}\frac{\partial\zeta_{32m}}{\partial\varphi}+\frac{\partial\zeta_{22m}}{\partial\varphi}\frac{\partial\zeta_{31m}}{\partial\varphi}\right.$$

$$\left.+\frac{\partial\zeta_{21m}}{\partial\nu}\frac{\partial\zeta_{32m}}{\partial\nu}+\frac{\partial\zeta_{22m}}{\partial\nu}\frac{\partial\zeta_{31m}}{\partial\nu}\right), \qquad (7.179)$$

$$\omega_{3m}^{(23)}=\zeta_{22m}^2u_{22m}^{(r)}+\zeta_{32m}^2u_{33m}^{(r)}$$

$$+\frac{u_{22m}^{(\varphi\nu)}}{s_{44m}}\left[\left(\frac{\partial\zeta_{22m}}{\partial\varphi}\right)^2+\left(\frac{\partial\zeta_{22m}}{\partial\nu}\right)^2\right]$$

$$+\frac{u_{33m}^{(\varphi\nu)}}{s_{44m}}\left[\left(\frac{\partial\zeta_{32m}}{\partial\varphi}\right)^2+\left(\frac{\partial\zeta_{32m}}{\partial\nu}\right)^2\right]$$

$$+\frac{u_{23m}^{(\varphi\nu)}}{s_{44m}}\left[\frac{\partial\zeta_{22m}}{\partial\varphi}\frac{\partial\zeta_{32m}}{\partial\varphi}+\frac{\partial\zeta_{22m}}{\partial\nu}\frac{\partial\zeta_{32m}}{\partial\nu}\right], \qquad (7.180)$$

$$\omega_{v}^{(23)}=\zeta_{2v}^2u_{22m}^{(r)}+\zeta_{3v}^2u_{33m}^{(r)}+\frac{u_{22m}^{(\varphi\nu)}}{s_{44m}}\left[\left(\frac{\partial\zeta_{2v}}{\partial\varphi}\right)^2+\left(\frac{\partial\zeta_{2v}}{\partial\nu}\right)^2\right]$$

$$+\frac{u_{33m}^{(\varphi\nu)}}{s_{44m}}\left[\left(\frac{\partial\zeta_{3v}}{\partial\varphi}\right)^2+\left(\frac{\partial\zeta_{3v}}{\partial\nu}\right)^2\right]+\frac{u_{23m}^{(\varphi\nu)}}{s_{44m}}\left[\frac{\partial\zeta_{2v}}{\partial\varphi}\frac{\partial\zeta_{3v}}{\partial\varphi}+\frac{\partial\zeta_{2v}}{\partial\nu}\frac{\partial\zeta_{3v}}{\partial\nu}\right],$$

(7.181)

where s_{44m} is given by Equation (3.35).

With regard to Equations (3.47), (7.72), the elastic energy W_m, which is accumulated in the cell matrix I, is given by Equation (7.78). With regard to Equations (7.72), (7.178)–(7.180), the coefficient $\Omega_{im}^{(12)}$ $(i=1,2,3)$, which is included in Equation (7.78), is replaced by $\Omega_{im}^{(23)}$ $(i=1,2,3)$. The coefficient $\Omega_{im}^{(23)}$ $(i=1,2,3)$ has the form

$$\Omega_{1m}^{(23)} = \int_0^{\pi/2}\int_0^{\pi/2}\int_{R_p}^{r_c} \omega_{1m}^{(23)}\, r^2\, dr\, d\varphi\, d\nu$$
$$= \int_0^{\pi/2}\int_0^{\pi/2} \left\{ \zeta_{21m}^2 U_{22m}^{(r)} + \zeta_{31m}^2 U_{33m}^{(r)} \right.$$
$$+ \frac{U_{22m}^{(\varphi\nu)}}{s_{44m}}\left[\left(\frac{\partial\zeta_{21m}}{\partial\varphi}\right)^2 + \left(\frac{\partial\zeta_{21m}}{\partial\nu}\right)^2\right]$$
$$+ \frac{U_{33m}^{(\varphi\nu)}}{s_{44m}}\left[\left(\frac{\partial\zeta_{31m}}{\partial\varphi}\right)^2 + \left(\frac{\partial\zeta_{31m}}{\partial\nu}\right)^2\right]$$
$$\left. + \frac{U_{23m}^{(\varphi\nu)}}{s_{44m}}\left[\frac{\partial\zeta_{21m}}{\partial\varphi}\frac{\partial\zeta_{31m}}{\partial\varphi} + \frac{\partial\zeta_{21m}}{\partial\nu}\frac{\partial\zeta_{31m}}{\partial\nu}\right]\right\} d\varphi\, d\nu, \quad (7.182)$$

$$\Omega_{2m}^{(23)} = \int_0^{\pi/2}\int_0^{\pi/2}\int_{R_p}^{r_c} \omega_{2m}^{(23)}\, r^2\, dr\, d\varphi\, d\nu$$
$$= \int_0^{\pi/2}\int_0^{\pi/2} \left\{ 2\left[\zeta_{21m}\zeta_{22m} U_{22m}^{(r)} + \zeta_{31m}\zeta_{32m} U_{33m}^{(r)}\right]\right.$$
$$+ \frac{2U_{22m}^{(\varphi\nu)}}{s_{44m}}\left(\frac{\partial\zeta_{21m}}{\partial\varphi}\frac{\partial\zeta_{22m}}{\partial\varphi} + \frac{\partial\zeta_{21m}}{\partial\nu}\frac{\partial\zeta_{22m}}{\partial\nu}\right)$$
$$+ \frac{2U_{33m}^{(\varphi\nu)}}{s_{44m}}\left(\frac{\partial\zeta_{31m}}{\partial\varphi}\frac{\partial\zeta_{32m}}{\partial\varphi} + \frac{\partial\zeta_{31m}}{\partial\nu}\frac{\partial\zeta_{32m}}{\partial\nu}\right)$$

$$+ \frac{U_{23m}^{(\varphi\nu)}}{s_{44m}} \left(\frac{\partial \zeta_{21m}}{\partial \varphi} \frac{\partial \zeta_{32m}}{\partial \varphi} + \frac{\partial \zeta_{22m}}{\partial \varphi} \frac{\partial \zeta_{31m}}{\partial \varphi} \right)$$

$$+ \frac{U_{23m}^{(\varphi\nu)}}{s_{44m}} \left(\frac{\partial \zeta_{21m}}{\partial \nu} \frac{\partial \zeta_{32m}}{\partial \nu} + \frac{\partial \zeta_{22m}}{\partial \nu} \frac{\partial \zeta_{31m}}{\partial \nu} \right) \Bigg\} d\varphi\, d\nu, \quad (7.183)$$

$$\Omega_{3m}^{(23)} = \int_0^{\pi/2} \int_0^{\pi/2} \int_{R_p}^{r_c} \omega_{3m}^{(23)} r^2\, dr\, d\varphi\, d\nu$$

$$= \int_0^{\pi/2} \int_0^{\pi/2} \Bigg\{ \zeta_{22m}^2 U_{22m}^{(r)} + \zeta_{32m}^2 U_{33m}^{(r)}$$

$$+ \frac{U_{22m}^{(\varphi\nu)}}{s_{44m}} \left[\left(\frac{\partial \zeta_{22m}}{\partial \varphi} \right)^2 + \left(\frac{\partial \zeta_{22m}}{\partial \nu} \right)^2 \right]$$

$$+ \frac{U_{33m}^{(\varphi\nu)}}{s_{44m}} \left[\left(\frac{\partial \zeta_{32m}}{\partial \varphi} \right)^2 + \left(\frac{\partial \zeta_{32m}}{\partial \nu} \right)^2 \right]$$

$$+ \frac{U_{23m}^{(\varphi\nu)}}{s_{44m}} \left[\frac{\partial \zeta_{22m}}{\partial \varphi} \frac{\partial \zeta_{32m}}{\partial \varphi} + \frac{\partial \zeta_{22m}}{\partial \nu} \frac{\partial \zeta_{32m}}{\partial \nu} \right] \Bigg\} d\varphi\, d\nu, \quad (7.184)$$

where $U_{ijm}^{(x)}$ $(i, j = 2{,}3;\ x = r, \varphi\nu)$, $U_{iim}^{(r)}$, $U_{iim}^{(\varphi\nu)}$ are given by Equations (7.82), (7.85), (7.88), (7.138), (7.140). With regard to Equations (7.24), (7.30), we get

$$U_{23m}^{(\varphi\nu)} = \frac{r_c^{c_{3m}+1} - R_p^{c_{3m}+1}}{c_{3m}+1}, \quad (7.185)$$

where r_c, c_{im} $(i = 1{,}3)$ are given by Equations (3.15)–(3.18), (3.40), respectively.

With regard to Equations (3.48), (7.73), the elastic energy W_v, which is accumulated in the cell matrix *II*, is given by Equation (7.89). With regard to Equations (7.73), (7.181), the coefficient $\Omega_v^{(12)}$, which is included in Equation (7.89), is replaced by $\Omega_v^{(23)}$. The coefficient $\Omega_v^{(23)}$ has the form

$$\Omega_v^{(23)} = \int_0^{\pi/2} \int_0^{\pi/2} \int_{R_p}^{r_c} \omega_v^{(23)} r^2\, dr\, d\varphi\, d\nu$$

$$= \int_0^{\pi/2}\int_0^{\pi/2} \left\{ \zeta_{2v}^2 U_{22v}^{(r)} + \zeta_{3v}^2 U_{33v}^{(r)} \right.$$

$$+ \frac{U_{22v}^{(\varphi\nu)}}{s_{44m}} \left[\left(\frac{\partial \zeta_{2v}}{\partial \varphi} \right)^2 + \left(\frac{\partial \zeta_{2v}}{\partial \nu} \right)^2 \right]$$

$$+ \frac{U_{33v}^{(\varphi\nu)}}{s_{44m}} \left[\left(\frac{\partial \zeta_{3v}}{\partial \varphi} \right)^2 + \left(\frac{\partial \zeta_{3v}}{\partial \nu} \right)^2 \right]$$

$$\left. + \frac{U_{23v}^{(\varphi\nu)}}{s_{44m}} \left[\frac{\partial \zeta_{2v}}{\partial \varphi} \frac{\partial \zeta_{3v}}{\partial \varphi} + \frac{\partial \zeta_{2v}}{\partial \nu} \frac{\partial \zeta_{3v}}{\partial \nu} \right] \right\} d\varphi \, d\nu. \quad (7.186)$$

With regard to Equations (7.82), (7.91), $U_{ijv}^{(\varphi\nu)}$ $(i, j = 2,3;\ x = r,\ \varphi\nu)$ is given by Equations (7.85), (7.88), (7.138), (7.140), (7.185). In case of $U_{ijv}^{(x)}$, the radius R_p, which is included in Equations (7.85), (7.88), (7.138), (7.140), (7.185), is replaced by R_v.

With regard to Equations (5.13), (6.81), (7.78), (7.89), the displacement u_{1c0} along the axis x_i $(i = 1,2,3)$ on the cubic cell surface *I*, *II*, i.e., for $r = d/2$ (see Figure 2.1a), is given by Equations (6.89), (7.92), (7.93).

The coefficients $\Omega_{1m}^{(12)}, \Omega_{2m}^{(12)}, \Omega_{v}^{(12)}$, which are included in Equations (7.92), (7.93), are replaced by $\Omega_{1m}^{(23)}$, $\Omega_{2m}^{(23)}$, $\Omega_{v}^{(23)}$ (see Equations (7.182), (7.183), (7.186)), respectively.

7.5. Integration Constants $C_{1q} \neq 0, C_{2q} \neq 0, C_{3q} \neq 0$

7.5.1. Cell Matrix around Particle

As presented above (see page 55), $u_{1rm} = u_{1rm}(r)$, $u_{2rm} = u_{2rm}(r)$ (see Equations (7.5), (7.6)) are decreasing functions of the variable r due to $\partial u_{1rm}/\partial r = -[(2/3) + \ln r] < 0$ and $c_{3m} < 0$ (see Equation (3.40)), respectively, and u_{3rm} is a constant function with respect to the variable r (see Equations (3.40), (7.5)–(7.7)).

As presented in Section 5.2, the absolute values $|u'_{1m}|$, $\left|\varepsilon'_{ijm}\right|$, $\left|\sigma'_{ijm}\right|$ $(i, j = 1,2,3)$ are required to represent decreasing functions of $r \in \langle R_p, r_c \rangle$ (see Equations (3.19)–(3.23), (3.36)–(3.39)). These decreasing functions exhibit maximum values on the particle-matrix boundary, i.e., for $r = R_p$. These decreasing courses are ensured by the integration constants $C_{1m} \neq 0$, $C_{2m} \neq 0$, $C_{3m} \neq 0$, which are determined by the mandatory and additional boundary conditions (5.3), (5.4) and (5.7), respectively. With regard to Equations (5.3), (5.4), (5.7), (7.4)–(7.7), (7.11), (7.15), we get

$$C_{im} = \frac{1}{\xi_{im}\zeta_m}\left[p_p\,\zeta_m^{(i1)} + \frac{u'_{1c}\,\zeta_m^{(i2)}}{r_c}\right] = \frac{1}{\xi_{im}\zeta_m}\left[p_p\,\zeta_m^{(i1)} + \frac{u_{1c0}\,\zeta_m^{(i2)}}{d}\right], \quad i = 1,2,3, \tag{7.187}$$

where $u'_{1c}/r_c = u_{1c0}/d$ (see Equations (3.15)–(3.18), (5.4)); and r_c, ξ_{im} ($i = 1,2,3$) are given by Equations (3.15)–(3.18), (7.8)–(7.10), respectively. The radial stress p_p, which acts at the particle-matrix boundary, i.e., for $r = R_p$ (see Figure 2.1a), is determined in Section 7.5.3. The displacement u_{1c0} along the axis x_i ($i = 1,2,3$) on the cell surface *I*, *II* (see Equation (5.4)), i.e., for $r = d/2$ (see Figure 2.1), is determined in Section 7.5.4. The coefficients $\zeta_m^{(i1)}$, $\zeta_m^{(i2)}$ ($i = 1,2,3$), ζ_m have the forms

$$\zeta_m^{(11)} = c_{3m}\left(r_c + u'_{1c}\right)^{c_{3m}-1}, \tag{7.188}$$

$$\zeta_m^{(21)} = \frac{2}{3} + \ln\left(r_c + u'_{1c}\right), \tag{7.189}$$

$$\zeta_m^{(31)} = -r_c^{c_{3m}}\left[\frac{2}{3} + \ln\left(r_c + u'_{1c}\right) + c_{3m}\left(\frac{1}{3} - \ln r_c\right)\left(\frac{r_c + u'_{1c}}{r_c}\right)^{c_{3m}-1}\right], \tag{7.190}$$

$$\zeta_m^{(12)} = -\frac{2c_{2m}c_{3m}r_c}{R_p}\left(r_c + u'_{1c}\right)^{c_{3m}-1}, \tag{7.191}$$

$$\zeta_m^{(22)} = -\frac{2c_{2m}r_c}{R_p}\left[\frac{2}{3} + \ln\left(r_c + u'_{1c}\right)\right] \tag{7.192}$$

$$\zeta_m^{(32)} = r_c R_p^{c_{3m}-1}\left\{c_{3m}\left[\frac{2\left(c_{1m} + 2c_{2m}\right)}{3} + \left(c_{1m} - c_{2m}\right)\ln R_p\right]\left(\frac{r_c + u'_{1c}}{R_p}\right)^{c_{3m}-1}\right.$$

$$- \left[\left(c_{1m} + c_{2m} \right) c_{3m} - 2 c_{2m} \right] \left[\frac{2}{3} + \ln \left(r_c + u'_{1c} \right) \right] \Bigg\}, \quad (7.193)$$

$$\zeta_m = c_{3m} \left(r_c + u'_{1c} \right)^{c_{3m}-1} \times \left\{ \left[\frac{2 \left(c_{1m} + 2 c_{2m} \right)}{3} + \left(c_{1m} - c_{2m} \right) \ln R_p \right] - \frac{2 c_{2m} r_c}{R_p} \left(\frac{1}{3} - \ln r_c \right) \right\} - \left\{ \left[\left(c_{1m} + c_{2m} \right) c_{3m} - 2 c_{2m} \right] R_p^{c_{3m}-1} + \frac{2 c_{2m} r_c^{c_{3m}}}{R_p} \right\} \left[\frac{2}{3} + \ln \left(r_c + u'_{1c} \right) \right], \quad (7.194)$$

where c_{im} $(i = 1,2,3)$ is given by Equation (3.40). With regard to Equations (7.4)–(7.7), (7.11)–(7.18), (7.187), we get

$$u'_{1m} = \frac{1}{\zeta_m} \left\{ p_p \left[\zeta_m^{(11)} r \left(\frac{1}{3} - \ln r \right) + \zeta_m^{(21)} r^{c_{3m}} + \zeta_m^{(31)} \right] + \frac{u_{1c0}}{d} \left[\zeta_m^{(12)} r \left(\frac{1}{3} - \ln r \right) + \zeta_m^{(22)} r^{c_{3m}} + \zeta_m^{(32)} \right] \right\}, \quad (7.195)$$

$$\varepsilon'_{11m} = -\frac{1}{\zeta_m} \left\{ p_p \left[\zeta_m^{(11)} \left(\frac{2}{3} + \ln r \right) - \zeta_m^{(21)} c_{3m} r^{c_{3m}-1} \right] + \frac{u_{1c0}}{d} \left[\zeta_m^{(12)} \left(\frac{2}{3} + \ln r \right) - \zeta_m^{(22)} c_{3m} r^{c_{3m}-1} \right] \right\}, \quad (7.196)$$

$$\varepsilon'_{22m} = \varepsilon'_{33m} = \frac{1}{\zeta_m} \left\{ p_p \left[\zeta_m^{(11)} \left(\frac{1}{3} - \ln r \right) + \zeta_m^{(21)} r^{c_{3m}-1} + \frac{\zeta_m^{(31)}}{r} \right] + \frac{u_{1c0}}{d} \left[\zeta_m^{(12)} \left(\frac{1}{3} - \ln r \right) + \zeta_m^{(22)} r^{c_{3m}-1} + \frac{\zeta_m^{(32)}}{r} \right] \right\}, \quad (7.197)$$

$$\varepsilon'_{12m} = \frac{1}{\zeta_m} \left(\left(\frac{1}{3} - \ln r \right) \left[\zeta_m^{(11)} \frac{\partial p_p}{\partial \varphi} + p_p \frac{\partial \zeta_m^{(11)}}{\partial \varphi} - \frac{p_p \zeta_m^{(11)}}{\zeta_m} \frac{\partial \zeta_m^{(11)}}{\partial \varphi} \right] \right.$$

$$+ r^{c_{3m}-1}\left[\zeta_m^{(21)}\frac{\partial p_p}{\partial\varphi} + p_p\frac{\partial\zeta_m^{(21)}}{\partial\varphi} - \frac{p_p\,\zeta_m^{(21)}}{\zeta_m}\frac{\partial\zeta_m^{(21)}}{\partial\varphi}\right]$$

$$+ \frac{1}{r}\left[\zeta_m^{(31)}\frac{\partial p_p}{\partial\varphi} + p_p\frac{\partial\zeta_m^{(31)}}{\partial\varphi} - \frac{p_p\,\zeta_m^{(31)}}{\zeta_m}\frac{\partial\zeta_m^{(31)}}{\partial\varphi}\right]$$

$$+ \frac{u_{1c0}}{d}\left\{\left(\frac{1}{3}-\ln r\right)\left[\frac{\partial\zeta_m^{(12)}}{\partial\varphi} - \frac{\zeta_m^{(12)}}{\zeta_m}\frac{\partial\zeta_m}{\partial\varphi}\right]\right.$$

$$+ r^{c_{3m}-1}\left[\frac{\partial\zeta_m^{(22)}}{\partial\varphi} - \frac{\zeta_m^{(22)}}{\zeta_m}\frac{\partial\zeta_m}{\partial\varphi}\right]$$

$$\left.\left. + \frac{1}{r}\left[\frac{\partial\zeta_m^{(32)}}{\partial\varphi} - \frac{\zeta_m^{(32)}}{\zeta_m}\frac{\partial\zeta_m}{\partial\varphi}\right]\right\}\right), \tag{7.198}$$

$$\varepsilon'_{13m} = \frac{1}{\zeta_m}\left(\left(\frac{1}{3}-\ln r\right)\left[\zeta_m^{(11)}\frac{\partial p_p}{\partial\nu} + p_p\frac{\partial\zeta_m^{(11)}}{\partial\nu} - \frac{p_p\,\zeta_m^{(11)}}{\zeta_m}\frac{\partial\zeta_m^{(11)}}{\partial\nu}\right]\right.$$

$$+ r^{c_{3m}-1}\left[\zeta_m^{(21)}\frac{\partial p_p}{\partial\nu} + p_p\frac{\partial\zeta_m^{(21)}}{\partial\nu} - \frac{p_p\,\zeta_m^{(21)}}{\zeta_m}\frac{\partial\zeta_m^{(21)}}{\partial\nu}\right]$$

$$+ \frac{1}{r}\left[\zeta_m^{(31)}\frac{\partial p_p}{\partial\nu} + p_p\frac{\partial\zeta_m^{(31)}}{\partial\nu} - \frac{p_p\,\zeta_m^{(31)}}{\zeta_m}\frac{\partial\zeta_m^{(31)}}{\partial\nu}\right]$$

$$+ \frac{u_{1c0}}{d}\left\{\left(\frac{1}{3}-\ln r\right)\left[\frac{\partial\zeta_m^{(12)}}{\partial\nu} - \frac{\zeta_m^{(12)}}{\zeta_m}\frac{\partial\zeta_m}{\partial\nu}\right]\right.$$

$$+ r^{c_{3m}-1}\left[\frac{\partial\zeta_m^{(22)}}{\partial\nu} - \frac{\zeta_m^{(22)}}{\zeta_m}\frac{\partial\zeta_m}{\partial\nu}\right]$$

$$\left.\left. + \frac{1}{r}\left[\frac{\partial\zeta_m^{(32)}}{\partial\nu} - \frac{\zeta_m^{(32)}}{\zeta_m}\frac{\partial\zeta_m}{\partial\nu}\right]\right\}\right), \tag{7.199}$$

$$\sigma'_{11m} = -\frac{1}{\zeta_m}\left(p_p\left\{\zeta_m^{(11)}\left[\frac{2\,(c_{1m}+2\,c_{2m})}{3} + (c_{1m}-c_{2m})\ln r\right]\right.\right.$$

$$\left. - \zeta_m^{(21)}\left[(c_{1m}+c_{2m})\,c_{3m} - 2\,c_{2m}\right] r^{c_{3m}-1} + \frac{2\,c_{2m}\,\zeta_m^{(31)}}{r}\right\}$$

$$+\frac{u_{1c0}}{d}\left\{\zeta_m^{(12)}\left[\frac{2\left(c_{1m}+2c_{2m}\right)}{3}+\left(c_{1m}-c_{2m}\right)\ln r\right]\right.$$
$$\left.\left.-\zeta_m^{(22)}\left[\left(c_{1m}+c_{2m}\right)c_{3m}-2c_{2m}\right]r^{c_{3m}-1}+\frac{2c_{2m}\zeta_m^{(32)}}{r}\right\}\right), \tag{7.200}$$

$$\sigma'_{22m}=\sigma'_{33m}=\frac{1}{\zeta_m}\left(p_p\left\{\zeta_m^{(11)}\left[\frac{c_{1m}+2c_{2m}}{3}-\left(c_{1m}-c_{2m}\right)\ln r\right]\right.\right.$$
$$\left.+\zeta_m^{(21)}\left(c_{1m}-c_{2m}c_{3m}\right)r^{c_{3m}-1}+\frac{c_{1m}\zeta_m^{(31)}}{r}\right\}$$
$$+\frac{u_{1c0}}{d}\left\{\zeta_m^{(12)}\left[\frac{c_{1m}+2c_{2m}}{3}-\left(c_{1m}-c_{2m}\right)\ln r\right]\right.$$
$$\left.\left.+\zeta_m^{(22)}\left(c_{1m}-c_{2m}c_{3m}\right)r^{c_{3m}-1}+\frac{c_{1m}\zeta_m^{(32)}}{r}\right\}\right), \tag{7.201}$$

$$\sigma'_{12m}=\frac{1}{s_{44m}\zeta_m}\left(\left(\frac{1}{3}-\ln r\right)\left[\zeta_m^{(11)}\frac{\partial p_p}{\partial\varphi}+p_p\frac{\partial\zeta_m^{(11)}}{\partial\varphi}-\frac{p_p\zeta_m^{(11)}}{\zeta_m}\frac{\partial\zeta_m^{(11)}}{\partial\varphi}\right]\right.$$
$$+r^{c_{3m}-1}\left[\zeta_m^{(21)}\frac{\partial p_p}{\partial\varphi}+p_p\frac{\partial\zeta_m^{(21)}}{\partial\varphi}-\frac{p_p\zeta_m^{(21)}}{\zeta_m}\frac{\partial\zeta_m^{(21)}}{\partial\varphi}\right]$$
$$+\frac{1}{r}\left[\zeta_m^{(31)}\frac{\partial p_p}{\partial\varphi}+p_p\frac{\partial\zeta_m^{(31)}}{\partial\varphi}-\frac{p_p\zeta_m^{(31)}}{\zeta_m}\frac{\partial\zeta_m^{(31)}}{\partial\varphi}\right]$$
$$+\frac{u_{1c0}}{d}\left\{\left(\frac{1}{3}-\ln r\right)\left[\frac{\partial\zeta_m^{(12)}}{\partial\varphi}-\frac{\zeta_m^{(12)}}{\zeta_m}\frac{\partial\zeta_m}{\partial\varphi}\right]\right.$$
$$+r^{c_{3m}-1}\left[\frac{\partial\zeta_m^{(22)}}{\partial\varphi}-\frac{\zeta_m^{(22)}}{\zeta_m}\frac{\partial\zeta_m}{\partial\varphi}\right]$$
$$\left.\left.+\frac{1}{r}\left[\frac{\partial\zeta_m^{(32)}}{\partial\varphi}-\frac{\zeta_m^{(32)}}{\zeta_m}\frac{\partial\zeta_m}{\partial\varphi}\right]\right\}\right), \tag{7.202}$$

$$\sigma'_{13m} = \frac{1}{s_{44m}\zeta_m}\Bigg(\left(\frac{1}{3} - \ln r\right)\left[\zeta_m^{(11)}\frac{\partial p_p}{\partial \nu} + p_p\frac{\partial \zeta_m^{(11)}}{\partial \nu} - \frac{p_p\zeta_m^{(11)}}{\zeta_m}\frac{\partial \zeta_m^{(11)}}{\partial \nu}\right]$$
$$+ r^{c_{3m}-1}\left[\zeta_m^{(21)}\frac{\partial p_p}{\partial \nu} + p_p\frac{\partial \zeta_m^{(21)}}{\partial \nu} - \frac{p_p\zeta_m^{(21)}}{\zeta_m}\frac{\partial \zeta_m^{(21)}}{\partial \nu}\right]$$
$$+ \frac{1}{r}\left[\zeta_m^{(31)}\frac{\partial p_p}{\partial \nu} + p_p\frac{\partial \zeta_m^{(31)}}{\partial \nu} - \frac{p_p\zeta_m^{(31)}}{\zeta_m}\frac{\partial \zeta_m^{(31)}}{\partial \nu}\right]$$
$$+ \frac{u_{1c0}}{d}\Bigg\{\left(\frac{1}{3} - \ln r\right)\left[\frac{\partial \zeta_m^{(12)}}{\partial \nu} - \frac{\zeta_m^{(12)}}{\zeta_m}\frac{\partial \zeta_m}{\partial \nu}\right]$$
$$+ r^{c_{3m}-1}\left[\frac{\partial \zeta_m^{(22)}}{\partial \nu} - \frac{\zeta_m^{(22)}}{\zeta_m}\frac{\partial \zeta_m}{\partial \nu}\right]$$
$$+ \frac{1}{r}\left[\frac{\partial \zeta_m^{(32)}}{\partial \nu} - \frac{\zeta_m^{(32)}}{\zeta_m}\frac{\partial \zeta_m}{\partial \nu}\right]\Bigg\}\Bigg), \tag{7.203}$$

where s_{44m} is given by Equation (3.35).

7.5.2. Cell Matrix around Void

As presented in Section 5.3, the absolute values $|u'_{1\nu}|$, $\left|\varepsilon'_{ij\nu}\right|$, $\left|\sigma'_{ij\nu}\right|$ $(i, j = 1,2,3)$ are required to represent increasing functions of $r \in \langle R_\nu, r_c \rangle$ (see Equations (3.19)–(3.23), (3.36)–(3.39)). These increasing functions exhibit maximum values on the cell surface, i.e., for $r = r_c$ (see Equations (3.15)–(3.18)). These increasing courses are ensured by the integration constants $C_{1\nu} \neq 0$, $C_{2\nu} \neq 0$, $C_{3\nu} \neq 0$, which are determined by the mandatory and additional boundary conditions (5.5), (5.6) and (5.8), respectively. With regard to Equations (5.5), (5.6), (5.8), (7.4)–(7.7), (7.15), we get

$$C_{i\nu} = -\frac{u'_{1c}\chi_{i\nu}}{\xi_{i\nu}\chi_\nu r_c} = -\frac{u_{1c0}\chi_{i\nu}}{d\,\xi_{i\nu}\chi_\nu}, \quad i = 1,2,3, \tag{7.204}$$

where $u'_{1c}/r_c = u_{1c0}/d$ (see Equations (3.15)–(3.18), (5.4)); and r_c, $\xi_{i\nu}$ $(i = 1,2,3)$ are given by Equations (3.15)–(3.18), (7.8)–(7.10), respectively. The displacement u_{1c0} along the axis x_i $(i = 1,2,3)$ on the cell surface I, II (see Equation (5.4)), i.e., for $r = d/2$ (see Figure 2.1), is determined in Section 7.5.4. The coefficients χ_{iv} $(i = 1,2,3)$, χ_v, have the forms

$$\chi_{1v} = -\frac{2c_{2m}c_{3m}r_c}{R_v}\left(r_c - u'_{1c}\right)^{c_{3m}-1}, \tag{7.205}$$

$$\chi_{2v} = -\frac{2c_{2m}r_c}{R_v}\left[\frac{2}{3} + \ln\left(r_c - u'_{1c}\right)\right] \tag{7.206}$$

$$\chi_{3v} = r_c R_v^{c_{3m}-1}\left\{c_{3m}\left[\frac{2\left(c_{1m}+2c_{2m}\right)}{3} + \left(c_{1m}-c_{2m}\right)\ln R_v\right]\left(\frac{r_c - u'_{1c}}{R_v}\right)^{c_{3m}-1} - \left[\left(c_{1m}+c_{2m}\right)c_{3m} - 2c_{2m}\right]\left[\frac{2}{3} + \ln\left(r_c - u'_{1c}\right)\right]\right\}, \tag{7.207}$$

$$\chi_v = c_{3m}\left(r_c - u'_{1c}\right)^{c_{3m}-1} \times\left\{\left[\frac{2\left(c_{1m}+2c_{2m}\right)}{3} + \left(c_{1m}-c_{2m}\right)\ln R_v\right] - \frac{2c_{2m}r_c}{R_v}\left(\frac{1}{3} - \ln r_c\right)\right\} - \left\{\left[\left(c_{1m}+c_{2m}\right)c_{3m} - 2c_{2m}\right]R_v^{c_{3m}-1} + \frac{2c_{2m}r_c^{c_{3m}}}{R_v}\right\}\left[\frac{2}{3} + \ln\left(r_c - u'_{1c}\right)\right], \tag{7.208}$$

where c_{im} $(i = 1,2,3)$ is given by Equation (3.40). With regard to Equations (7.4)–(7.7), (7.11)–(7.18), (7.204), we get

$$u'_{1v} = -\frac{u_{1c0}}{d\chi_v}\left[\chi_{1v}r\left(\frac{1}{3} - \ln r\right) + \chi_{2v}r^{c_{3m}} + \chi_{3v}\right], \tag{7.209}$$

$$\varepsilon'_{11v} = \frac{u_{1c0}}{d\chi_v}\left[\chi_{1v}r\left(\frac{2}{3} + \ln r\right) - \chi_{2v}c_{3m}r^{c_{3m}-1}\right], \tag{7.210}$$

$$\varepsilon'_{22v} = \varepsilon'_{33v} = -\frac{u_{1c0}}{d\chi_v}\left[\chi_{1v}\left(\frac{1}{3} - \ln r\right) + \chi_{2v}r^{c_{3m}-1} + \frac{\chi_{3v}}{r}\right], \tag{7.211}$$

$$\varepsilon'_{12v} = -\frac{u_{1c0}}{d\chi_v}\left\{\left(\frac{1}{3} - \ln r\right)\left[\frac{\partial\chi_{1v}}{\partial\varphi} - \frac{\chi_{1v}}{\chi_v}\frac{\partial\chi_v}{\partial\varphi}\right]\right.$$

$$+ r^{c_{3m}-1}\left[\frac{\partial \chi_{2v}}{\partial \varphi} - \frac{\chi_{2v}}{\chi_v}\frac{\partial \chi_v}{\partial \varphi}\right] + \frac{1}{r}\left[\frac{\partial \chi_{3v}}{\partial \varphi} - \frac{\chi_{3v}}{\chi_v}\frac{\partial \chi_v}{\partial \varphi}\right]\Bigg\}, \quad (7.212)$$

$$\varepsilon'_{13v} = -\frac{u_{1c0}}{d\chi_v}\Bigg\{\left(\frac{1}{3} - \ln r\right)\left[\frac{\partial \chi_{1v}}{\partial \nu} - \frac{\chi_{1v}}{\chi_v}\frac{\partial \chi_v}{\partial \nu}\right] + r^{c_{3m}-1}\left[\frac{\partial \chi_{2v}}{\partial \nu} - \frac{\chi_{2v}}{\chi_v}\frac{\partial \chi_v}{\partial \nu}\right] + \frac{1}{r}\left[\frac{\partial \chi_{3v}}{\partial \nu} - \frac{\chi_{3v}}{\chi_v}\frac{\partial \chi_v}{\partial \nu}\right]\Bigg\}, \quad (7.213)$$

$$\sigma'_{11v} = \frac{u_{1c0}}{d\chi_v}\Bigg\{\chi_{1v}\left[\frac{2\left(c_{1m} + 2c_{2m}\right)}{3} + \left(c_{1m} - c_{2m}\right)\ln r\right] - \chi_{2v}\left[\left(c_{1m} + c_{2m}\right)c_{3m} - 2c_{2m}\right]r^{c_{3m}-1} + \frac{2c_{2m}\chi_{3v}}{r}\Bigg\}, \quad (7.214)$$

$$\sigma'_{22v} = \sigma'_{33v} = -\frac{u_{1c0}}{d\chi_v}\Bigg\{\chi_{1v}\left[\frac{c_{1m} + 2c_{2m}}{3} - \left(c_{1m} - c_{2m}\right)\ln r\right] + \chi_{2v}\left(c_{1m} - c_{2m}c_{3m}\right)r^{c_{3m}-1} + \frac{c_{1m}\chi_{3v}}{r}\Bigg\}, \quad (7.215)$$

$$\sigma'_{12v} - -\frac{u_{1c0}}{d\,s_{44m}\chi_v}\Bigg\{\left(\frac{1}{3} - \ln r\right)\left[\frac{\partial \chi_{1v}}{\partial \varphi} - \frac{\chi_{1v}}{\chi_v}\frac{\partial \chi_v}{\partial \varphi}\right] + r^{c_{3m}-1}\left[\frac{\partial \chi_{2v}}{\partial \varphi} - \frac{\chi_{2v}}{\chi_v}\frac{\partial \chi_v}{\partial \varphi}\right] + \frac{1}{r}\left[\frac{\partial \chi_{3v}}{\partial \varphi} - \frac{\chi_{3v}}{\chi_v}\frac{\partial \chi_v}{\partial \varphi}\right]\Bigg\}, \quad (7.216)$$

$$\sigma'_{13v} = -\frac{u_{1c0}}{d\,s_{44m}\chi_v}\Bigg\{\left(\frac{1}{3} - \ln r\right)\left[\frac{\partial \chi_{1v}}{\partial \nu} - \frac{\chi_{1v}}{\chi_v}\frac{\partial \chi_v}{\partial \nu}\right]$$

$$+ r^{c_{3m}-1}\left[\frac{\partial \chi_{2v}}{\partial v} - \frac{\chi_{2v}}{\chi_v}\frac{\partial \chi_v}{\partial v}\right]$$
$$+ \frac{1}{r}\left[\frac{\partial \chi_{3v}}{\partial v} - \frac{\chi_{3v}}{\chi_v}\frac{\partial \chi_v}{\partial v}\right]\Bigg\}, \tag{7.217}$$

where s_{44m} is given by Equation (3.35).

7.5.3. Radial Stress p_p

With regard to Equations (5.12), (6.25), (7.197), the radial stress p_p, which acts at the particle-matrix boundary, i.e., for $r = R_p$ (see Figure 2.1a), has the form

$$p_p = \zeta_{pm}\left(\frac{u_{1c0}\,\zeta_{2m}}{d} + \beta_m - \beta_p\right), \tag{7.218}$$

where $u'_{1c}/r_c = u_{1c0}/d$ (see Equations (3.15)–(3.18), (5.4)). The coefficients ζ_{pm}, ζ_{1m}, ζ_{2m} are derived as

$$\zeta_{pm} = \frac{\xi_{3p}}{\xi_{3p}\,\zeta_{1m} - 1}, \tag{7.219}$$

$$\zeta_{1m} = \frac{1}{\zeta_m}\left[\zeta_m^{(11)}\left(\frac{1}{3} - \ln R_p\right) + \zeta_m^{(21)} R_p^{c_{3m}-1} + \frac{\zeta_m^{(31)}}{R_p}\right], \tag{7.220}$$

$$\zeta_{2m} = \frac{1}{\zeta_m}\left[\zeta_m^{(12)}\left(\frac{1}{3} - \ln R_p\right) + \zeta_m^{(22)} R_p^{c_{3m}-1} + \frac{\zeta_m^{(32)}}{R_p}\right], \tag{7.221}$$

where c_{3m}, ξ_{3p}, ζ_m, $\zeta_m^{(i1)}$, $\zeta_m^{(i2)}$ $(i = 1,2,3)$ are given by Equations (3.40), (6.19), (7.188)–(7.194), respectively.

7.5.4. Displacement u_{1c0}

With regard to Equations (7.187), (7.204), (7.218), the integration constant C_{iq} $(i = 1,2,3;\ q = m,v)$ has the form

$$C_{im} = \frac{1}{\xi_{im}}\left[\frac{u_{1c0}\,\zeta_{i1m}}{d} + \zeta_{i2m}\left(\beta_m - \beta_p\right)\right], \quad i = 1,2,3, \tag{7.222}$$

$$C_{iv} = \frac{u_{1c0}\zeta_{iv}}{d\xi_{iv}}, \quad i = 1,2,3, \tag{7.223}$$

where ξ_{im}, ξ_{im} ($i = 1,2,3$) are given by Equations (7.8)–(7.10), and β_q ($q = p,m$) is determined in Section 4.2. The coefficients ζ_{i1m}, ζ_{i2m}, ζ_{iv} ($i = 1,2,3$) are derived as

$$\zeta_{i1m} = \frac{1}{\zeta_m}\left[\zeta_m^{(i1)}\zeta_{pm}\zeta_{2m} + \zeta_m^{(i2)}\right], \quad i = 1,2,3, \tag{7.224}$$

$$\zeta_{i2m} = \frac{\zeta_m^{(i1)}\zeta_{pm}}{\zeta_m}, \quad i = 1,2,3, \tag{7.225}$$

$$\zeta_{iv} = -\frac{\chi_{iv}}{\chi_v}, \quad i = 1,2,3, \tag{7.226}$$

where ζ_m, $\zeta_m^{(i1)}$, $\zeta_m^{(i2)}$, χ_v, χ_{iv} ($i = 1,2,3$), ζ_{pm}, ζ_{2m} are given by Equations (7.188)–(7.194), (7.205)–(7.208), (7.219), (7.221), respectively.

The elastic energy W_m, which is accumulated in the cell matrix *I*, is given by Equation (7.78). The coefficient $\Omega_{im}^{(12)}$ ($i = 1,2,3$), which is included in Equation (7.78), is replaced by $\Omega_{im}^{(123)} = \Omega_{im}^{(12)} + \Omega_{im}^{(13)} + \Omega_{im}^{(23)}$, where $\Omega_{im}^{(12)}$, $\Omega_{im}^{(13)}$, $\Omega_{im}^{(23)}$ are given by Equations (7.79)–(7.81), (7.134)–(7.136), (7.182)–(7.184), respectively. The coefficients ζ_{i1m}, ζ_{i2m} ($i = 1,2,3$), which are included in Equations (7.79)–(7.81), (7.134)–(7.136), (7.182)–(7.184), are given by Equations (7.224), (7.225), respectively.

The elastic energy W_v, which is accumulated in the cell matrix *II*, is given by Equation (7.89). The coefficient $\Omega_v^{(12)}$ ($i = 1,2,3$), which is included in Equation (7.89), is replaced by $\Omega_v^{(123)} = \Omega_{imv}^{(12)} + \Omega_v^{(13)} + \Omega_v^{(23)}$, where $\Omega_v^{(12)}$, $\Omega_v^{(13)}$, $\Omega_v^{(23)}$ are given by Equations (7.90), (7.141), (7.186), respectively. The coefficient ζ_{iv} ($i = 1,2,3$), which is included in Equations (7.90), (7.141), (7.186), is given by Equation (7.226).

With regard to Equations (7.188)–(7.194), (7.205)–(7.208), the elastic energy $W = W_p + W_m - W_v$ (see Equation (5.13)) is not a polynomial function of the variable u_{1c0}. The condition (5.13) determines a minimum of the function $W = W(u_{1c0})$. This minimum is determined for a real two-component material by a numerical method. Finally, such a numerical value of the variable u_{1c0} is determined, which results in a minimum numerical value of $W = W_p + W_m - W_v$.

Chapter 8

Analytical Model 3

The mathematical procedure in Section 8.1, which is applied to the differential equation (3.41), (3.42), results in the differential equation (8.3) with a non-zero right-hand side. Consequently, Wronskian's method is applied to Equation (8.3) [25]. The solution of Equation (8.3) includes the integration constants C_1, C_2, C_3 (see Equation (8.4)).

With regard to Equations (8.11)–(8.18), ε'_{ijq}, σ'_{ijq} (see Equations (3.19)–(3.23), (3.36)–(3.39)) are functions of $f_1 = \ln r$, $f_1 = r^{c_{3q}-1}$, $f_3 = 1/r$, where $c_{3q} < 0$ (see Equation (3.40)), and then we get $\left(\varepsilon'_{ijq}\right)_{r \to 0} \to \pm\infty$, $\left(\sigma'_{ijq}\right)_{r \to 0} \to \pm\infty$. Consequently, we get $C_{1p} = C_{2p} = C_{3p} = 0$, and the thermal and phase-transformation-induced stresses and strains in the spherical particle are determined in Section 6.2.

In case of the cell matrix *I, II* (see Figure 2.1a), the following solutions for the displacement $u'_{1q} = u'_{1q}(r, \varphi, \nu)$ $(q = m, v)$ are obtained:

1. the solutions (8.37), (8.51), which consider the integration constants C_{1q}, C_{2q} $(q = m, v)$, where $C_{3q} = 0$ (see Equations (8.32), (8.33), (8.46), (8.47)). These integration constants are determined by the mathematical boundary conditions (5.3)–(5.6).

2. the solutions (8.99), (8.113), which consider the integration constants C_{1q}, C_{3q} $(q = m, v)$, where $C_{2q} = 0$ (see Equations (8.94), (8.95), (8.108), (8.109)). These integration constants are determined by the mathematical boundary conditions (5.3)–(5.6).

3. the solutions (8.149), (8.163), which consider the integration constants C_{2q}, C_{3q} $(q = m,v)$, where $C_{1q} = 0$ (see Equations (8.144), (8.145), (8.158), (8.159)). These integration constants are determined by the mathematical boundary conditions (5.3)–(5.6).

4. the solution (8.200), (8.214), which consider the integration constants C_{1q}, C_{2q}, C_{3q} $(q = m,v)$ (see Equations (8.192), (8.209)). These integration constants are determined by the mathematical boundary conditions (5.3)–(5.8).

8.1. Mathematical Procedure 3

Performing ∂^2Eq.$(3.41)/\partial r^2$, the differential equation (3.41) is transformed to the form

$$r\frac{\partial^3 u_1'}{\partial r^3} + 4r^2\frac{\partial^2 u_1'}{\partial r^2} + \frac{r}{s_{44}(c_1 + c_2)}\frac{\partial U_1'}{\partial r} = 0, \tag{8.1}$$

where the coefficients c_1, c_2 and the function $U_1' = U_1'(r,\varphi,\nu)$ are given by Equations (3.40) and (3.43), respectively. Considering the right-hand sides of Equations (3.42), (7.2), we get

$$r\frac{\partial U_1'}{\partial r} = c_3\left(C_1 r + C_2 r^{c_3} + C_3\right), \tag{8.2}$$

where c_3 is given by Equation (3.40). The integration constant C_i $(i = 1,2,3)$ is determined by the boundary conditions, which are defined in Chapter 5.

Substituting Equation (8.2) to Equation (8.1), the differential equation (8.1) is transformed to the form

$$\frac{\partial^3 u_1'}{\partial r^3} + \frac{4}{r}\frac{\partial^2 u_1'}{\partial r^2} = -\frac{c_3}{s_{44}(c_1 + c_2)}\left(\frac{C_1}{r^2} + C_2 r^{c_3 - 3} + \frac{C_3}{r^3}\right), \tag{8.3}$$

where the elastic modulus s_{44} is given by Equation (3.35).

Equation (8.3) with the variable r represents a linear differential equation of the third order with respect to u_1'. Due to the non-zero right-hand side, Wronskian's method is applied to Equation (8.3) [25]. Consequently, the radial displacement $u_{1q}' = u_{1q}'(r,\varphi,\nu)$ $(q = m,v)$ has the form

$$u'_{1q} = \sum_{i=1}^{2} C_{iq}\,\xi_{iq}\,u_{irq}, \quad q = m, v. \tag{8.4}$$

The subscripts $q = m$ and $q = v$ are related to the cell matrix *I* and *II* (see Figure 2.1a), where $r \in \langle R_p, r_c \rangle$ and $r \in \langle R_v, r_c \rangle$, respectively. The function $u_{irq} = u_{irq}(r)$ ($i = 1,2,3$; $q = m,v$) is derived as

$$u_{1rq} = r\left(\frac{4}{3} - \ln r\right), \tag{8.5}$$

$$u_{2rq} = r^{c_{3q}}, \quad c_{3q} < 0, \tag{8.6}$$

$$u_{3rq} = \frac{1}{2} + \ln r, \tag{8.7}$$

where $u_{2rq} = u_{2rq}(r)$ is a decreasing function of the variable r due to $c_{3q} < 0$ (see Equation (3.40)), and $u_{3rq} = u_{3rq}(r)$ is an increasing function of the variable r. The function $u_{1rq} = u_{1rq}(r)$ increases on the condition $\partial u_{1rq}/\partial r = (1/3) - \ln r > 0$. This condition results in $r < e^{1/3} \approx 1.39561$ m, where $e \approx 2.71828$ is Euler's number [25]. In case of real two-component material, we get $R_q < 1.39561$ m ($q = p,v$; see Figure 2.1a). With regard to Equation (3.35), (3.40), the coefficient ξ_{iq} ($i = 1,2,3$; $q = m,v$) has the form

$$\xi_{1q} = \frac{c_3}{3s_{44q}(c_{1q} + c_{2q})} = \frac{2(2\mu_q - 1)}{3}, \tag{8.8}$$

$$\xi_{2q} = 3\xi_{1q}\left[\frac{1}{2c_{3q}} - \frac{1}{3(c_{3q} - 1)} - \frac{1}{6(c_{3q} + 2)}\right] = \frac{1}{4(1 - \mu_q)(4\mu_q - 5)}, \tag{8.9}$$

$$\xi_{3q} = \frac{3\xi_{1q}}{2} = 2\mu_q - 1. \tag{8.10}$$

With regard to Equations (3.19)–(3.23), (3.36)–(3.39), (3.45), (8.4)–(8.7), we get

$$\varepsilon'_{11q} = C_{1q}\xi_{1q}\left(\frac{1}{3} - \ln r\right) + C_{2q}\xi_{2q}c_{3q}r^{c_{3q} - 1} + \frac{C_{3q}\xi_{3q}}{r}, \tag{8.11}$$

$$\varepsilon'_{22q} = \varepsilon'_{33q} = C_{1q}\xi_{1q}\left(\frac{4}{3} - \ln r\right) + C_{2q}\xi_{2q}r^{c_{3q}-1} + \frac{C_{3q}\xi_{3q}}{r}\left(\frac{1}{2} + \ln r\right), \quad (8.12)$$

$$\varepsilon'_{12q} = \xi_{1q}\left(\frac{4}{3} - \ln r\right)\frac{\partial C_{1q}}{\partial \varphi} + \xi_{2q}r^{c_{3q}-1}\frac{\partial C_{2q}}{\partial \varphi} + \frac{\xi_{3q}}{r}\left(\frac{1}{2} + \ln r\right)\frac{\partial C_{3q}}{\partial \varphi}, \quad (8.13)$$

$$\varepsilon'_{13q} = \xi_{1q}\left(\frac{4}{3} - \ln r\right)\frac{\partial C_{1q}}{\partial \nu} + \xi_{2q}r^{c_{3q}-1}\frac{\partial C_{2q}}{\partial \nu} + \frac{\xi_{3q}}{r}\left(\frac{1}{2} + \ln r\right)\frac{\partial C_{3q}}{\partial \nu}, \quad (8.14)$$

$$\begin{aligned}\sigma'_{11q} = {} & C_{1q}\xi_{1q}\left[\frac{c_{1q} - 7c_{2q}}{3} - (c_{1q} - c_{2q})\ln r\right] \\ & + C_{2q}\xi_{2q}\left[(c_{1q} + c_{2q})c_{3q} - 2c_{2q}\right]r^{c_{3q}-1} \\ & + \frac{C_{3q}\xi_{3q}}{r}(c_{1q} - 2c_{2q}\ln r),\end{aligned} \quad (8.15)$$

$$\begin{aligned}\sigma'_{22q} = \sigma'_{33q} = {} & C_{1q}\xi_{1q}\left[\frac{4c_{1q} - c_{2q}}{3} - (c_{1q} - c_{2q})\ln r\right] \\ & + C_{2q}\xi_{2q}(c_{1q} - c_{2q}c_{3q})r^{c_{3q}-1} \\ & + \frac{C_{3q}\xi_{3q}}{r}\left(\frac{c_{1q} - 2c_{2q}}{2} + c_{1q}\ln r\right),\end{aligned} \quad (8.16)$$

$$\sigma'_{12q} = \frac{1}{s_{44q}}\left[\xi_{1q}\left(\frac{4}{3} - \ln r\right)\frac{\partial C_{1q}}{\partial \varphi} + \xi_{2q}r^{c_{3q}-1}\frac{\partial C_{2q}}{\partial \varphi} + \frac{\xi_{3q}}{r}\left(\frac{1}{2} + \ln r\right)\frac{\partial C_{3q}}{\partial \varphi},\right], \quad (8.17)$$

$$\sigma'_{13q} = \frac{1}{s_{44q}}\left[\xi_{1q}\left(\frac{4}{3} - \ln r\right)\frac{\partial C_{1q}}{\partial \nu} + \xi_{2q}r^{c_{3q}-1}\frac{\partial C_{2q}}{\partial \nu} + \frac{\xi_{3q}}{r}\left(\frac{1}{2} + \ln r\right)\frac{\partial C_{3q}}{\partial \nu},\right], \quad (8.18)$$

$$w_q = \sum_{i,j=1;\, i\le j}^{3} \xi_{iq}\xi_{jq}\left[C_{iq}C_{jq}u_{ijq}^{(r)} + \frac{u_{ijq}^{(\varphi\nu)}}{s_{44m}}\left(\frac{\partial C_i}{\partial\varphi}\frac{\partial C_j}{\partial\varphi} + \frac{\partial C_i}{\partial\nu}\frac{\partial C_j}{\partial\nu}\right)\right], \tag{8.19}$$

where the functions $u_{ijq}^{(r)} = u_{ijq}^{(r)}(r)$, $u_{ijq}^{(\varphi\nu)} = u_{ijq}^{(\varphi\nu)}(r)$ $(i, j=1,2,3;\ i\le j)$ are derived as

$$u_{11q}^{(r)} = \frac{c_{2q}-c_{1q}}{2}\ln^2 r + \frac{c_{1q}-c_{2q}}{3}\ln r + \frac{17c_{1q}+c_{2q}}{18}, \tag{8.20}$$

$$u_{12q}^{(r)} = c_{3q}(c_{1q}-c_{2q})\, r^{c_{3q}-1}\ln r + \left[2c_{1q} + \frac{c_{3q}(c_{2q}-7c_{1q})}{3}\right] r^{c_{3q}-1}, \tag{8.21}$$

$$u_{13q}^{(r)} = (3c_{1q}-c_{2q})\frac{\ln r}{r} - \frac{4c_{1q}-c_{2q}}{3r}, \tag{8.22}$$

$$u_{22q}^{(r)} = \left[\frac{c_{3q}^2(c_{1q}+c_{2q})}{2} + c_{1q}(1-2c_{3q})\right] r^{2(c_{3q}-1)}, \tag{8.23}$$

$$u_{23q}^{(r)} = 2c_{1q}(1-c_{3q})\, r^{c_{3q}-2}\ln r + (c_{2q}c_{3q}-c_{1q})\, r^{c_{3q}-2}, \tag{8.24}$$

$$u_{33q}^{(r)} = \frac{c_{1q}\ln^2 r}{r^2} - \frac{c_{1q}\ln r}{r^2} + \frac{c_{2q}-2c_{1q}}{4r^2}, \tag{8.25}$$

$$u_{11q}^{(\varphi\nu)} = \ln^2 r - \frac{8}{3}\ln r + \frac{16}{9}, \tag{8.26}$$

$$u_{12q}^{(\varphi\nu)} = \frac{8}{3} r^{c_{3q}-1} - 2r^{c_{3q}-1}\ln r, \tag{8.27}$$

$$u_{13q}^{(\varphi\nu)} = \frac{4}{3r} + \frac{5\ln r}{3r} - \frac{2\ln^2 r}{r}, \tag{8.28}$$

$$u_{22q}^{(\varphi\nu)} = r^{2(c_{3q}-1)}, \tag{8.29}$$

$$u_{23q}^{(\varphi\nu)} = 2r^{c_{3q}-2}\ln r + r^{c_{3q}-2}, \tag{8.30}$$

$$u_{33q}^{(\varphi\nu)} = \frac{\ln^2 r}{r^2} + \frac{\ln r}{r^2} + \frac{1}{4r^2}. \tag{8.31}$$

8.2. Integration Constants $C_{1q} \neq 0, C_{2q} \neq 0, C_{3q} = 0$

As presented above (see page 103), the integration constant C_{1q}, C_{2q}, C_{3q} ($q = m,v$) (see Equation (8.4)) are determined with respect to the following conditions

- $C_{1q} \neq 0, C_{2q} \neq 0, C_{3q} = 0,$
- $C_{1q} \neq 0, C_{3q} \neq 0, C_{2q} = 0,$
- $C_{1q} = 0, C_{2q} \neq 0, C_{3q} \neq 0,$
- $C_{1q} \neq 0, C_{2q} \neq 0, C_{3q} \neq 0.$

8.2.1. Cell Matrix around Particle

As presented above (see page 105), $u_{1rm} = u_{1rm}(r)$ is an increasing function of the variable r for $r < e^{1/3} \approx 1.39561$ m (see Equation (8.5)), and $u_{2rm} = u_{2rm}(r)$ is a decreasing function of the variable r due to $c_{3m} < 0$ (see Equations (3.40), (8.6)).

As presented in Section 5.2, the absolute values $|u'_{1m}|$, $\left|\varepsilon'_{ijm}\right|$, $\left|\sigma'_{ijm}\right|$ ($i, j = 1,2,3$) are required to represent decreasing functions of $r \in \langle R_p, r_c \rangle$ (see Equations (3.19)–(3.23), (3.36)–(3.39)). These decreasing functions exhibit maximum values on the particle-matrix boundary, i.e., for $r = R_p$. These decreasing courses are ensured by the integration constants $C_{1m} \neq 0$, $C_{2m} \neq 0$, which are determined by the mandatory boundary conditions (5.3), (5.4), where $C_{3m} = 0$. With regard to Equations (5.3), (5.4), (8.4)–(8.6), (8.15), we get

$$C_{1m} = -\frac{1}{\xi_{1m}\zeta_m}\left(p_p + \frac{u'_{1c}\zeta_{1m}}{r_c}\right) = -\frac{1}{\xi_{1m}\zeta_m}\left(p_p + \frac{u_{1c0}\zeta_{1m}}{d}\right), \quad (8.32)$$

$$\begin{aligned} C_{2m} &= \frac{1}{\xi_{2m}\zeta_m r_c^{c_{3m}-1}}\left[p_p\left(\frac{4}{3} - \ln r_c\right) + \frac{u'_{1c}\zeta_{2m}}{r_c}\right] \\ &= \frac{1}{\xi_{2m}\zeta_m r_c^{c_{3m}-1}}\left[p_p\left(\frac{4}{3} - \ln r_c\right) + \frac{u_{1c0}\zeta_{2m}}{d}\right], \end{aligned} \quad (8.33)$$

where $u'_{1c}/r_c = u_{1c0}/d$ (see Equations (3.15)–(3.18), (5.4)); and r_c, ξ_{im} ($i = 1,2$) are given by Equations (3.15)–(3.18), (8.8), (8.9), respectively. The radial stress p_p, which acts at the particle-matrix boundary, i.e., for $r = R_p$ (see Figure 2.1a), is determined in Section 8.2.3. The displacement u_{1c0} along the axis x_j $(j=1,2,3)$ on the cell surface *I*, *II* (see Equation (5.4)), i.e., for $r = d/2$ (see Figure 2.1), is determined in Section 8.2.4. The coefficients ζ_{im} $(i=1,2)$, ζ_m have the forms

$$\zeta_{1m} = [(c_{1m} + c_{2m})\, c_{3m} - 2\, c_{2m}] \left(\frac{R_p}{r_c}\right)^{c_{3m}-1}, \tag{8.34}$$

$$\zeta_{2m} = \frac{c_{1m} - 7c_{2m}}{3} - (c_{1m} - c_{2m}) \ln R_p, \tag{8.35}$$

$$\zeta_m = \zeta_{2m} - \zeta_{1m} \left(\frac{4}{3} - \ln r_c\right), \tag{8.36}$$

where c_{im} $(i=1,2,3)$ is given by Equation (3.40). With regard to Equations (8.4)–(8.6), (8.11)–(8.18), (8.32), (8.33), we get

$$u'_{1m} = -\frac{r}{\zeta_m} \left\{ p_p \left[\frac{4}{3} - \ln r - \left(\frac{4}{3} - \ln r_c\right) \left(\frac{r}{r_c}\right)^{c_{3m}-1} \right] + \frac{u_{1c0}}{d} \left[\zeta_{1m} \left(\frac{4}{3} - \ln r\right) - \zeta_{2m} \left(\frac{r}{r_c}\right)^{c_{3m}-1} \right] \right\}, \tag{8.37}$$

$$\varepsilon'_{11m} = -\frac{1}{\zeta_m} \left\{ p_p \left[\frac{1}{3} - \ln r - c_{3m} \left(\frac{4}{3} - \ln r_c\right) \left(\frac{r}{r_c}\right)^{c_{3m}-1} \right] + \frac{u_{1c0}}{d} \left[\zeta_{1m} \left(\frac{1}{3} - \ln r\right) - c_{3m} \zeta_{2m} \left(\frac{r}{r_c}\right)^{c_{3m}-1} \right] \right\}, \tag{8.38}$$

$$\varepsilon'_{22m} = \varepsilon'_{33m} = -\frac{1}{\zeta_m} \left\{ p_p \left[\frac{4}{3} - \ln r - \left(\frac{4}{3} - \ln r_c\right) \left(\frac{r}{r_c}\right)^{c_{3m}-1} \right] + \frac{u_{1c0}}{d} \left[\zeta_{1m} \left(\frac{4}{3} - \ln r\right) - \zeta_{2m} \left(\frac{r}{r_c}\right)^{c_{3m}-1} \right] \right\}, \tag{8.39}$$

$$\varepsilon'_{12m} = -\frac{1}{\zeta_m}\left(\left[\frac{4}{3} - \ln r - \left(\frac{4}{3} - \ln r_c\right)\left(\frac{r}{r_c}\right)^{c_{3m}-1}\right]\left(\frac{\partial p_p}{\partial \varphi} - \frac{p_p}{\zeta_m}\frac{\partial \zeta_m}{\partial \varphi}\right)\right.$$
$$+ \frac{p_p}{r_c}\left[1 - (1 - c_{3m})\left(\frac{4}{3} - \ln r_c\right)\right]\left(\frac{r}{r_c}\right)^{c_{3m}-1}\frac{\partial r_c}{\partial \varphi}$$
$$- \frac{u_{1c0}}{d\,\zeta_m}\left[\zeta_{1m}\left(\frac{4}{3} - \ln r\right) - \zeta_{2m}\left(\frac{r}{r_c}\right)^{c_{3m}-1}\right]\frac{\partial \zeta_m}{\partial \varphi}$$
$$+ \frac{u_{1c0}}{d}\left[\left(\frac{4}{3} - \ln r\right)\frac{\partial \zeta_{1m}}{\partial \varphi} - \left(\frac{r}{r_c}\right)^{c_{3m}-1}\frac{\partial \zeta_{2m}}{\partial \varphi}\right.$$
$$\left.\left. - \frac{\zeta_{2m}(1 - c_{3m})}{r_c}\left(\frac{r}{r_c}\right)^{c_{3m}-1}\frac{\partial r_c}{\partial \varphi}\right]\right), \quad (8.40)$$

$$\varepsilon'_{13m} = -\frac{1}{\zeta_m}\left(\left[\frac{4}{3} - \ln r - \left(\frac{4}{3} - \ln r_c\right)\left(\frac{r}{r_c}\right)^{c_{3m}-1}\right]\left(\frac{\partial p_p}{\partial \nu} - \frac{p_p}{\zeta_m}\frac{\partial \zeta_m}{\partial \nu}\right)\right.$$
$$+ \frac{p_p}{r_c}\left[1 - (1 - c_{3m})\left(\frac{4}{3} - \ln r_c\right)\right]\left(\frac{r}{r_c}\right)^{c_{3m}-1}\frac{\partial r_c}{\partial \nu}$$
$$- \frac{u_{1c0}}{d\,\zeta_m}\left[\zeta_{1m}\left(\frac{4}{3} - \ln r\right) - \zeta_{2m}\left(\frac{r}{r_c}\right)^{c_{3m}-1}\right]\frac{\partial \zeta_m}{\partial \nu}$$
$$+ \frac{u_{1c0}}{d}\left[\left(\frac{4}{3} - \ln r\right)\frac{\partial \zeta_{1m}}{\partial \nu} - \left(\frac{r}{r_c}\right)^{c_{3m}-1}\frac{\partial \zeta_{2m}}{\partial \nu}\right.$$
$$\left.\left. - \frac{\zeta_{2m}(1 - c_{3m})}{r_c}\left(\frac{r}{r_c}\right)^{c_{3m}-1}\frac{\partial r_c}{\partial \nu}\right]\right), \quad (8.41)$$

$$\sigma'_{11m} = \frac{1}{\zeta_m}\left(p_p\left\{\frac{7c_{2m} - c_{1m}}{3} + (c_{1m} - c_{2m})\ln r\right.\right.$$
$$\left. + [(c_{1m} + c_{2m})\,c_{3m} - 2\,c_{2m}]\left(\frac{4}{3} - \ln r_c\right)\left(\frac{r}{r_c}\right)^{c_{3m}-1}\right\}$$
$$+ \frac{u_{1c0}}{d}\left\{\zeta_{1m}\left[\frac{7c_{2m} - c_{1m}}{3} + (c_{1m} - c_{2m})\ln r\right]\right.$$

$$+\zeta_{2m}\left[(c_{1m}+c_{2m})c_{3m}-2c_{2m}\right]\left(\frac{r}{r_c}\right)^{c_{3m}-1}\Bigg\}\Bigg), \quad (8.42)$$

$$\sigma'_{22m}=\sigma'_{33m}=\frac{1}{\zeta_m}\Bigg(p_p\left[\frac{c_{2m}-4c_{1m}}{3}+(c_{1m}-c_{2m})\ln r\right.$$
$$\left.+(c_{1m}-c_{2m}c_{3m})\left(\frac{4}{3}-\ln r_c\right)\left(\frac{r}{r_c}\right)^{c_{3m}-1}\right]$$
$$+\frac{u_{1c0}}{d}\Bigg\{\zeta_{1m}\left[\frac{c_{2m}-4c_{1m}}{3}+(c_{1m}-c_{2m})\ln r\right]$$
$$+\zeta_{2m}(c_{1m}-c_{2m}c_{3m})\left(\frac{r}{r_c}\right)^{c_{3m}-1}\Bigg\}\Bigg), \quad (8.43)$$

$$\sigma'_{12m}=-\frac{1}{s_{44m}\zeta_m}\Bigg(\left[\frac{4}{3}-\ln r-\left(\frac{4}{3}-\ln r_c\right)\left(\frac{r}{r_c}\right)^{c_{3m}-1}\right]\left(\frac{\partial p_p}{\partial\varphi}-\frac{p_p}{\zeta_m}\frac{\partial\zeta_m}{\partial\varphi}\right)$$
$$+\frac{p_p}{r_c}\left[1-(1-c_{3m})\left(\frac{4}{3}-\ln r_c\right)\right]\left(\frac{r}{r_c}\right)^{c_{3m}-1}\frac{\partial r_c}{\partial\varphi}$$
$$-\frac{u_{1c0}}{d\,\zeta_m}\left[\zeta_{1m}\left(\frac{4}{3}-\ln r\right)-\zeta_{2m}\left(\frac{r}{r_c}\right)^{c_{3m}-1}\right]\frac{\partial\zeta_m}{\partial\varphi}$$
$$+\frac{u_{1c0}}{d}\Bigg[\left(\frac{4}{3}-\ln r\right)\frac{\partial\zeta_{1m}}{\partial\varphi}-\left(\frac{r}{r_c}\right)^{c_{3m}-1}\frac{\partial\zeta_{2m}}{\partial\varphi}$$
$$-\frac{\zeta_{2m}(1-c_{3m})}{r_c}\left(\frac{r}{r_c}\right)^{c_{3m}-1}\frac{\partial r_c}{\partial\varphi}\Bigg]\Bigg), \quad (8.44)$$

$$\sigma'_{13m}=-\frac{1}{s_{44m}\zeta_m}\Bigg(\left[\frac{4}{3}-\ln r-\left(\frac{4}{3}-\ln r_c\right)\left(\frac{r}{r_c}\right)^{c_{3m}-1}\right]\left(\frac{\partial p_p}{\partial\nu}-\frac{p_p}{\zeta_m}\frac{\partial\zeta_m}{\partial\nu}\right)$$
$$+\frac{p_p}{r_c}\left[1-(1-c_{3m})\left(\frac{4}{3}-\ln r_c\right)\right]\left(\frac{r}{r_c}\right)^{c_{3m}-1}\frac{\partial r_c}{\partial\nu}$$
$$-\frac{u_{1c0}}{d\,\zeta_m}\left[\zeta_{1m}\left(\frac{4}{3}-\ln r\right)-\zeta_{2m}\left(\frac{r}{r_c}\right)^{c_{3m}-1}\right]\frac{\partial\zeta_m}{\partial\nu}$$

$$+\frac{u_{1c0}}{d}\left[\left(\frac{4}{3}-\ln r\right)\frac{\partial\zeta_{1m}}{\partial\nu}-\left(\frac{r}{r_c}\right)^{c_{3m}-1}\frac{\partial\zeta_{2m}}{\partial\nu}\right.$$

$$\left.\left.-\frac{\zeta_{2m}\left(1-c_{3m}\right)}{r_c}\left(\frac{r}{r_c}\right)^{c_{3m}-1}\frac{\partial r_c}{\partial\nu}\right]\right), \quad (8.45)$$

where s_{44m} is given by Equation (3.35).

8.2.2. Cell Matrix around Void

As presented in Section 5.3, the absolute values $|u'_{1v}|$, $\left|\varepsilon'_{ijv}\right|$, $\left|\sigma'_{ijv}\right|$ $(i,j=1,2,3)$ are required to represent increasing functions of $r \in \langle R_v, r_c\rangle$ (see Equations (3.19)–(3.23), (3.36)–(3.39)). These increasing functions exhibit maximum values on the cell surface, i.e., for $r = r_c$ (see Equations (3.15)–(3.18)). These increasing courses are ensured by the integration constants $C_{1v} \neq 0$, $C_{2v} \neq 0$, which are determined by the mandatory boundary conditions (5.5), (5.6), where $C_{3v} = 0$. With regard to Equations (5.5), (5.6), (8.4)–(8.6), (8.15), we get

$$C_{1v}=\frac{u'_{1c}\chi_{1v}}{\xi_{1v}\chi_v r_c}=\frac{u_{1c0}\chi_{1v}}{d\,\xi_{1v}\chi_v}, \quad (8.46)$$

$$C_{2v}=-\frac{u'_{1c}\chi_{2v}}{\xi_{2v}\chi_v r_c^{c_{3m}}}=-\frac{u_{1c0}\chi_{2v}}{d\,\xi_{2v}\chi_v r_c^{c_{3m}-1}}, \quad (8.47)$$

where $u'_{1c}/r_c = u_{1c0}/d$ (see Equations (3.15)–(3.18), (5.4)); and r_c, ξ_{iv} $(i=1,2)$ are given by Equations (3.15)–(3.18), (8.8), (8.9), respectively. The displacement u_{1c0} along the axis x_j $(j=1,2,3)$ on the cell surface *I*, *II* (see Equation (5.4)), i.e., for $r = d/2$ (see Figure 2.1), is determined in Section 8.2.4. The coefficients χ_{iv} $(i=1,2)$, χ_v have the forms

$$\chi_{1v}=\left[\left(c_{1m}+c_{2m}\right)c_{3m}-2c_{2m}\right]\left(\frac{R_v}{r_c}\right)^{c_{3m}-1}, \quad (8.48)$$

$$\chi_{2v}=\frac{c_{1m}-7c_{2m}}{3}-\left(c_{1m}-c_{2m}\right)\ln R_v, \quad (8.49)$$

$$\chi_v=\chi_{2v}-\chi_{1v}\left(\frac{4}{3}-\ln r_c\right), \quad (8.50)$$

where c_{im} $(i=1,2,3)$ is given by Equation (3.40). With regard to Equations (8.4)–(8.6), (8.11)–(8.18), (8.46), (8.47), we get

$$u'_{1v} = \frac{r\,u_{1c0}}{d\chi_v}\left[\chi_{1v}\left(\frac{4}{3} - \ln r\right) - \chi_{2v}\left(\frac{r}{r_c}\right)^{c_{3m}-1}\right], \tag{8.51}$$

$$\varepsilon'_{11v} = \frac{u_{1c0}}{d\chi_v}\left[\chi_{1v}\left(\frac{1}{3} - \ln r\right) - c_{3m}\chi_{2v}\left(\frac{r}{r_c}\right)^{c_{3m}-1}\right], \tag{8.52}$$

$$\varepsilon'_{22v} = \varepsilon'_{33v} = \frac{u_{1c0}}{d\chi_v}\left[\chi_{1v}\left(\frac{4}{3} - \ln r\right) - \chi_{2v}\left(\frac{r}{r_c}\right)^{c_{3m}-1}\right], \tag{8.53}$$

$$\begin{aligned}\varepsilon'_{12v} = \frac{u_{1c0}}{d\chi_m}\Bigg\{&\left(\frac{4}{3} - \ln r\right)\frac{\partial\chi_{1m}}{\partial\varphi} - \left(\frac{r}{r_c}\right)^{c_{3m}-1}\frac{\partial\chi_{2m}}{\partial\varphi} \\ &- \frac{1}{\chi_m}\left[\chi_{1m}\left(\frac{4}{3} - \ln r\right) - \chi_{2m}\left(\frac{r}{r_c}\right)^{c_{3m}-1}\right]\frac{\partial\chi_m}{\partial\varphi} \\ &- \frac{\chi_{2m}(1-c_{3m})}{r_c}\left(\frac{r}{r_c}\right)^{c_{3m}-1}\frac{\partial r_c}{\partial\varphi}\Bigg\},\end{aligned} \tag{8.54}$$

$$\begin{aligned}\varepsilon'_{13v} = \frac{u_{1c0}}{d\chi_m}\Bigg\{&\left(\frac{4}{3} - \ln r\right)\frac{\partial\chi_{1m}}{\partial\nu} - \left(\frac{r}{r_c}\right)^{c_{3m}-1}\frac{\partial\chi_{2m}}{\partial\nu} \\ &- \frac{1}{\chi_m}\left[\chi_{1m}\left(\frac{4}{3} - \ln r\right) - \chi_{2m}\left(\frac{r}{r_c}\right)^{c_{3m}-1}\right]\frac{\partial\chi_m}{\partial\nu} \\ &- \frac{\chi_{2m}(1-c_{3m})}{r_c}\left(\frac{r}{r_c}\right)^{c_{3m}-1}\frac{\partial r_c}{\partial\nu}\Bigg\},\end{aligned} \tag{8.55}$$

$$\begin{aligned}\sigma'_{11v} = -\frac{u_{1c0}}{d\chi_v}\Bigg\{&\chi_{1m}\left[\frac{7c_{2m} - c_{1m}}{3} + (c_{1m} - c_{2m})\ln r\right] \\ &+ \chi_{2m}\left[(c_{1m} + c_{2m})c_{3m} - 2c_{2m}\right]\left(\frac{r}{r_c}\right)^{c_{3m}-1}\Bigg\},\end{aligned} \tag{8.56}$$

$$\sigma'_{22v} = \sigma'_{33m} = -\frac{u_{1c0}}{d\chi_v}\left\{\chi_{1m}\left[\frac{c_{2m}-4c_{1m}}{3} + (c_{1m}-c_{2m})\ln r\right] + \chi_{2m}(c_{1m}-c_{2m}c_{3m})\left(\frac{r}{r_c}\right)^{c_{3m}-1}\right\}, \quad (8.57)$$

$$\sigma'_{12v} = \frac{u_{1c0}}{d s_{44m}\chi_m}\left\{\left(\frac{4}{3}-\ln r\right)\frac{\partial\chi_{1m}}{\partial\varphi} - \left(\frac{r}{r_c}\right)^{c_{3m}-1}\frac{\partial\chi_{2m}}{\partial\varphi} - \frac{1}{\chi_m}\left[\chi_{1m}\left(\frac{4}{3}-\ln r\right) - \chi_{2m}\left(\frac{r}{r_c}\right)^{c_{3m}-1}\right]\frac{\partial\chi_m}{\partial\varphi} - \frac{\chi_{2m}(1-c_{3m})}{r_c}\left(\frac{r}{r_c}\right)^{c_{3m}-1}\frac{\partial r_c}{\partial\varphi}\right\}, \quad (8.58)$$

$$\sigma'_{13v} = \frac{u_{1c0}}{d s_{44m}\chi_m}\left\{\left(\frac{4}{3}-\ln r\right)\frac{\partial\chi_{1m}}{\partial\nu} - \left(\frac{r}{r_c}\right)^{c_{3m}-1}\frac{\partial\chi_{2m}}{\partial\nu} - \frac{1}{\chi_m}\left[\chi_{1m}\left(\frac{4}{3}-\ln r\right) - \chi_{2m}\left(\frac{r}{r_c}\right)^{c_{3m}-1}\right]\frac{\partial\chi_m}{\partial\nu} - \frac{\chi_{2m}(1-c_{3m})}{r_c}\left(\frac{r}{r_c}\right)^{c_{3m}-1}\frac{\partial r_c}{\partial\nu}\right\}, \quad (8.59)$$

where s_{44m} is given by Equation (3.35).

8.2.3. Radial Stress p_p

With regard to Equations (5.12), (6.25), (8.39), the radial stress p_p, which acts at the particle-matrix boundary, i.e., for $r = R_p$ (see Figure 2.1a), has the form

$$p_p = \zeta_{pm}\left(\frac{u_{1c0}\zeta_{4m}}{d} + \beta_m - \beta_p\right), \quad (8.60)$$

where $u'_{1c}/r_c = u_{1c0}/d$ (see Equations (3.15)–(3.18), (5.4)). The coefficients ζ_{pm}, ζ_{3m}, ζ_{4m} are derived as

$$\zeta_{pm} = \frac{\xi_{3p}}{1-\xi_{3p}\zeta_{3m}}, \tag{8.61}$$

$$\zeta_{3m} = \frac{1}{\zeta_m}\left[\zeta_{1m}\left(\frac{4}{3} - \ln R_p\right) - \left(\frac{4}{3} - \ln r_c\right)\left(\frac{R_p}{r_c}\right)^{c_{3m}-1}\right], \tag{8.62}$$

$$\zeta_{4m} = \frac{1}{\zeta_m}\left[\zeta_{1m}\left(\frac{4}{3} - \ln R_p\right) - \zeta_{2m}\left(\frac{R_p}{r_c}\right)^{c_{3m}-1}\right] \tag{8.63}$$

where r_c, c_{3m}, ξ_{3p}, ζ_m, ζ_{1m}, ζ_{2m} are given by Equations (3.15)–(3.18), (3.40), (6.19), (8.34)–(8.36), respectively.

8.2.4. Displacement u_{1c0}

With regard to Equations (8.32), (8.33), (8.46), (8.60), the integration constant C_{iq} ($i = 1,2$; $q = m,v$) has the form

$$C_{im} = \frac{1}{\xi_{im}}\left[\frac{u_{1c0}\zeta_{i1m}}{d} + \zeta_{i2m}(\beta_m - \beta_p)\right], \quad i = 1,2, \tag{8.64}$$

$$C_{iv} = \frac{u_{1c0}\zeta_{iv}}{d\xi_{iv}}, \quad i = 1,2, \tag{8.65}$$

where ξ_{im}, ξ_{iv} ($i = 1,2$) are given by Equations (8.8), (8.9), and β_q ($q = p,m$) is determined in Section 4.2. The coefficients ζ_{i1m}, ζ_{i2m}, ζ_{iv} ($i = 1,2$) are derived as

$$\zeta_{11m} = \frac{\zeta_{1m} + \zeta_{pm}\zeta_{4m}}{\zeta_m}, \tag{8.66}$$

$$\zeta_{21m} = \frac{1}{\zeta_m r^{c_{3m}-1}}\left[\zeta_{pm}\zeta_{4m}\left(\frac{4}{3} - \ln r_c\right) + \zeta_{2m}\right], \tag{8.67}$$

$$\zeta_{12m} = -\frac{\zeta_{pm}}{\zeta_m}, \tag{8.68}$$

$$\zeta_{22m} = \frac{\zeta_{pm}}{\zeta_m r^{c_{3m}-1}}\left(\frac{4}{3} - \ln r_c\right), \tag{8.69}$$

$$\zeta_{1v} = \frac{\chi_{1v}}{\chi_v}, \tag{8.70}$$

$$\zeta_{2v} = -\frac{\chi_{2v}}{\chi_v r^{c_{3m}-1}}, \tag{8.71}$$

where r_c, ζ_m, ζ_{im}, χ_v, χ_{iv} $(i = 1,2)$ and ζ_{pm}, ζ_{4m} are given by Equations (3.15)–(3.18), (8.34)–(8.36), (8.48)–(8.50) and (8.61), (8.63), respectively.

With regard to Equations (8.19)–(8.21), (8.23), (8.26), (8.27), (8.29), (8.64), (8.65), the elastic energy density w_q $(q = m,v)$ has the form

$$w_m = \omega_{1m}^{(12)} \left(\frac{u_{1c0}}{d}\right)^2 + \frac{\omega_{2m}^{(12)} u_{1c0} (\beta_m - \beta_p)}{d} + \omega_{3m}^{(12)} (\beta_m - \beta_p)^2, \tag{8.72}$$

$$w_v = \omega_v^{(12)} \left(\frac{u_{1c0}}{d}\right)^2, \tag{8.73}$$

where the coefficients $\omega_{im}^{(12)}$ $(i = 1,2,3)$, $\omega_v^{(12)}$ are derived as

$$\begin{aligned} \omega_{1m}^{(12)} &= \zeta_{11m}^2 u_{11m}^{(r)} + \zeta_{21m}^2 u_{22m}^{(r)} + \zeta_{11m} \zeta_{21m} u_{12m}^{(r)} \\ &+ \frac{u_{11m}^{(\varphi\nu)}}{s_{44m}} \left[\left(\frac{\partial \zeta_{11m}}{\partial \varphi}\right)^2 + \left(\frac{\partial \zeta_{11m}}{\partial \nu}\right)^2\right] \\ &+ \frac{u_{22m}^{(\varphi\nu)}}{s_{44m}} \left[\left(\frac{\partial \zeta_{21m}}{\partial \varphi}\right)^2 + \left(\frac{\partial \zeta_{21m}}{\partial \nu}\right)^2\right] \\ &+ \frac{u_{12m}^{(\varphi\nu)}}{s_{44m}} \left[\frac{\partial \zeta_{11m}}{\partial \varphi} \frac{\partial \zeta_{21m}}{\partial \varphi} + \frac{\partial \zeta_{11m}}{\partial \nu} \frac{\partial \zeta_{21m}}{\partial \nu}\right], \end{aligned} \tag{8.74}$$

$$\begin{aligned} \omega_{2m}^{(12)} &= 2 \left[\zeta_{11m} \zeta_{12m} u_{11m}^{(r)} + \zeta_{21m} \zeta_{22m} u_{22m}^{(r)}\right] + (\zeta_{11m} \zeta_{22m} + \zeta_{12m} \zeta_{21m})\, u_{12m}^{(r)} \\ &+ \frac{2 u_{11m}^{(\varphi\nu)}}{s_{44m}} \left(\frac{\partial \zeta_{11m}}{\partial \varphi} \frac{\partial \zeta_{12m}}{\partial \varphi} + \frac{\partial \zeta_{11m}}{\partial \nu} \frac{\partial \zeta_{12m}}{\partial \nu}\right) \\ &+ \frac{2 u_{22m}^{(\varphi\nu)}}{s_{44m}} \left(\frac{\partial \zeta_{21m}}{\partial \varphi} \frac{\partial \zeta_{22m}}{\partial \varphi} + \frac{\partial \zeta_{21m}}{\partial \nu} \frac{\partial \zeta_{22m}}{\partial \nu}\right) \\ &+ \frac{u_{12m}^{(\varphi\nu)}}{s_{44m}} \left(\frac{\partial \zeta_{11m}}{\partial \varphi} \frac{\partial \zeta_{22m}}{\partial \varphi} + \frac{\partial \zeta_{12m}}{\partial \varphi} \frac{\partial \zeta_{21m}}{\partial \varphi}\right. \end{aligned}$$

$$+\frac{\partial\zeta_{11m}}{\partial\nu}\frac{\partial\zeta_{22m}}{\partial\nu}+\frac{\partial\zeta_{12m}}{\partial\nu}\frac{\partial\zeta_{21m}}{\partial\nu}\Bigg), \tag{8.75}$$

$$\begin{aligned}\omega_{3m}^{(12)} &= \zeta_{12m}^2 u_{11m}^{(r)}+\zeta_{22m}^2 u_{22m}^{(r)}+\zeta_{12m}\zeta_{22m}u_{12m}^{(r)}\\ &+\frac{u_{11m}^{(\varphi\nu)}}{s_{44m}}\left[\left(\frac{\partial\zeta_{12m}}{\partial\varphi}\right)^2+\left(\frac{\partial\zeta_{12m}}{\partial\nu}\right)^2\right]\\ &+\frac{u_{22m}^{(\varphi\nu)}}{s_{44m}}\left[\left(\frac{\partial\zeta_{22m}}{\partial\varphi}\right)^2+\left(\frac{\partial\zeta_{22m}}{\partial\nu}\right)^2\right]\\ &+\frac{u_{12m}^{(\varphi\nu)}}{s_{44m}}\left[\frac{\partial\zeta_{12m}}{\partial\varphi}\frac{\partial\zeta_{22m}}{\partial\varphi}+\frac{\partial\zeta_{12m}}{\partial\nu}\frac{\partial\zeta_{22m}}{\partial\nu}\right],\end{aligned} \tag{8.76}$$

$$\begin{aligned}\omega_{v}^{(12)} &= \zeta_{1v}^2 u_{11m}^{(r)}+\zeta_{2v}^2 u_{22m}^{(r)}+\zeta_{1v}\zeta_{2v}u_{12m}^{(r)}+\frac{u_{11m}^{(\varphi\nu)}}{s_{44m}}\left[\left(\frac{\partial\zeta_{1v}}{\partial\varphi}\right)^2+\left(\frac{\partial\zeta_{1v}}{\partial\nu}\right)^2\right]\\ &+\frac{u_{22m}^{(\varphi\nu)}}{s_{44m}}\left[\left(\frac{\partial\zeta_{2v}}{\partial\varphi}\right)^2+\left(\frac{\partial\zeta_{2v}}{\partial\nu}\right)^2\right]+\frac{u_{12m}^{(\varphi\nu)}}{s_{44m}}\left[\frac{\partial\zeta_{1v}}{\partial\varphi}\frac{\partial\zeta_{2v}}{\partial\varphi}+\frac{\partial\zeta_{1v}}{\partial\nu}\frac{\partial\zeta_{2v}}{\partial\nu}\right],\end{aligned} \tag{8.77}$$

where s_{44m} is given by Equation (3.35).

With regard to Equations (3.47), (8.72), the elastic energy W_m, which is accumulated in the cell matrix I, has the form

$$W_m = 8\left[\Omega_{1m}^{(12)}\left(\frac{u_{1c0}}{d}\right)^2+\frac{\Omega_{2m}^{(12)}u_{1c0}(\beta_m-\beta_p)}{d}+\Omega_{3m}^{(12)}(\beta_m-\beta_p)^2\right]. \tag{8.78}$$

The coefficient $\Omega_{im}^{(12)}$ $(i=1,2,3)$ is derived as

$$\Omega_{1m}^{(12)} = \int_0^{\pi/2}\int_0^{\pi/2}\int_{R_p}^{r_c}\omega_{1m}^{(12)} r^2\,dr\,d\varphi\,d\nu$$

$$= \int_0^{\pi/2}\int_0^{\pi/2} \left\{ \zeta_{11m}^2 U_{11m}^{(r)} + \zeta_{21m}^2 U_{22m}^{(r)} + \zeta_{11m}\zeta_{21m} U_{12m}^{(r)} + \frac{U_{11m}^{(\varphi\nu)}}{s_{44m}} \left[\left(\frac{\partial \zeta_{11m}}{\partial \varphi}\right)^2 + \left(\frac{\partial \zeta_{11m}}{\partial \nu}\right)^2 \right] + \frac{U_{22m}^{(\varphi\nu)}}{s_{44m}} \left[\left(\frac{\partial \zeta_{21m}}{\partial \varphi}\right)^2 + \left(\frac{\partial \zeta_{21m}}{\partial \nu}\right)^2 \right] + \frac{U_{12m}^{(\varphi\nu)}}{s_{44m}} \left[\frac{\partial \zeta_{11m}}{\partial \varphi}\frac{\partial \zeta_{21m}}{\partial \varphi} + \frac{\partial \zeta_{11m}}{\partial \nu}\frac{\partial \zeta_{21m}}{\partial \nu} \right] \right\} d\varphi\, d\nu, \quad (8.79)$$

$$\Omega_{2m}^{(12)} = \int_0^{\pi/2}\int_0^{\pi/2}\int_{R_p}^{r_c} \omega_{2m}^{(12)} r^2\, dr\, d\varphi\, d\nu$$

$$= \int_0^{\pi/2}\int_0^{\pi/2} \left\{ 2\left[\zeta_{11m}\zeta_{12m} U_{11m}^{(r)} + \zeta_{21m}\zeta_{22m} U_{22m}^{(r)} \right] + \left(\zeta_{11m}\zeta_{22m} + \zeta_{12m}\zeta_{21m}\right) U_{12m}^{(r)} + \frac{2U_{11m}^{(\varphi\nu)}}{s_{44m}} \left(\frac{\partial \zeta_{11m}}{\partial \varphi}\frac{\partial \zeta_{12m}}{\partial \varphi} + \frac{\partial \zeta_{11m}}{\partial \nu}\frac{\partial \zeta_{12m}}{\partial \nu} \right) + \frac{2U_{22m}^{(\varphi\nu)}}{s_{44m}} \left(\frac{\partial \zeta_{21m}}{\partial \varphi}\frac{\partial \zeta_{22m}}{\partial \varphi} + \frac{\partial \zeta_{21m}}{\partial \nu}\frac{\partial \zeta_{22m}}{\partial \nu} \right) + \frac{U_{12m}^{(\varphi\nu)}}{s_{44m}} \left(\frac{\partial \zeta_{11m}}{\partial \varphi}\frac{\partial \zeta_{22m}}{\partial \varphi} + \frac{\partial \zeta_{12m}}{\partial \varphi}\frac{\partial \zeta_{21m}}{\partial \varphi} \right) + \frac{U_{12m}^{(\varphi\nu)}}{s_{44m}} \left(\frac{\partial \zeta_{11m}}{\partial \nu}\frac{\partial \zeta_{22m}}{\partial \nu} + \frac{\partial \zeta_{12m}}{\partial \nu}\frac{\partial \zeta_{21m}}{\partial \nu} \right) \right\} d\varphi\, d\nu, \quad (8.80)$$

$$\Omega_{3m}^{(12)} = \int_0^{\pi/2}\int_0^{\pi/2}\int_{R_p}^{r_c} \omega_{3m}^{(12)} r^2\, dr\, d\varphi\, d\nu$$

$$= \int_0^{\pi/2} \int_0^{\pi/2} \left\{ \zeta_{12m}^2 U_{11m}^{(r)} + \zeta_{22m}^2 U_{22m}^{(r)} + \zeta_{12m} \zeta_{22m} U_{12m}^{(r)} \right.$$
$$+ \frac{U_{11m}^{(\varphi\nu)}}{s_{44m}} \left[\left(\frac{\partial \zeta_{12m}}{\partial \varphi} \right)^2 + \left(\frac{\partial \zeta_{12m}}{\partial \nu} \right)^2 \right]$$
$$+ \frac{U_{22m}^{(\varphi\nu)}}{s_{44m}} \left[\left(\frac{\partial \zeta_{22m}}{\partial \varphi} \right)^2 + \left(\frac{\partial \zeta_{22m}}{\partial \nu} \right)^2 \right]$$
$$\left. + \frac{U_{12m}^{(\varphi\nu)}}{s_{44m}} \left[\frac{\partial \zeta_{12m}}{\partial \varphi} \frac{\partial \zeta_{22m}}{\partial \varphi} + \frac{\partial \zeta_{12m}}{\partial \nu} \frac{\partial \zeta_{22m}}{\partial \nu} \right] \right\} d\varphi \, d\nu, \quad (8.81)$$

where $U_{ijm}^{(x)}$ $(i, j = 1{,}2;\ x = r, \varphi\nu)$ has the form

$$U_{ijm}^{(x)} = \int_{R_p}^{r_c} u_{ijm}^{(x)} r^2 \, dr, \quad i, j = 1, 2, 3; \quad x = r,\ \varphi\nu. \quad (8.82)$$

With regard to Equations (8.20), (8.21), (8.23), (8.26), (8.27), (8.29), (8.82), we get

$$U_{11m}^{(r)} = \frac{c_{2m} - c_{1m}}{6} \left\{ r_c^3 \left[\left(\ln r_c - \frac{1}{3} \right)^2 + \frac{1}{9} \right] - R_p^3 \left[\left(\ln R_p - \frac{1}{3} \right)^2 + \frac{1}{9} \right] \right\}$$
$$+ \frac{c_{1m} - c_{2m}}{9} \left[r_c^3 \left(\ln r_c \quad \frac{1}{3} \right) \quad R_p^3 \left(\ln R_p \quad \frac{1}{3} \right) \right]$$
$$+ \frac{17\, c_{1m} + c_{2m}}{54} \left(r_c^3 - R_p^3 \right), \quad (8.83)$$

$$U_{12m}^{(r)} = \frac{c_{3m} \left(c_{1m} - c_{2m} \right)}{c_{3m} + 2}$$
$$\times \left[r_c^{c_{3m}+2} \left(\ln r_c - \frac{1}{c_{3m} + 2} \right) - R_p^{c_{3m}+2} \left(\ln R_p - \frac{1}{c_{3m} + 2} \right) \right]$$
$$+ \frac{1}{c_{3m} + 2} \left[2\, c_{1m} + \frac{c_{3m} \left(c_{2m} - 7 c_{1m} \right)}{3} \right] \left(r_c^{c_{3m}+2} - R_p^{c_{3m}+2} \right), \quad (8.84)$$

$$U_{22m}^{(r)} = \frac{1}{2\, c_{3m} + 1} \left[\frac{c_{3m}^2 \left(c_{1m} + c_{2m} \right)}{2} + c_{1m} \left(1 - 2\, c_{3m} \right) \right] \left(r_c^{2 c_{3m}+1} - R_p^{2 c_{3m}+1} \right),$$

(8.85)

$$U_{11m}^{(\varphi\nu)} = \frac{r_c^3}{3}\left[(\ln r_c - 3)\left(\ln r_c - \frac{1}{3}\right) + \frac{17}{9}\right] - \frac{R_p^3}{3}\left[(\ln R_p - 3)\left(\ln R_p - \frac{1}{3}\right) + \frac{17}{9}\right], \quad (8.86)$$

$$U_{12m}^{(\varphi\nu)} = \frac{2}{c_{3m}+2}\left\{r_c^{c_{3m}+2}\left[\frac{4c_{3m}+11}{3(c_{3m}+2)} - \ln r_c\right] - R_p^{c_{3m}+2}\left[\frac{4c_{3m}+11}{3(c_{3m}+2)} - \ln R_p\right]\right\}, \quad (8.87)$$

$$U_{22m}^{(\varphi\nu)} = \frac{r_c^{2c_{3m}+1} - R_p^{2c_{3m}+1}}{2c_{3m}+1}, \quad (8.88)$$

where r_c, c_{im} ($i = 1,2,3$) are given by Equations (3.15)–(3.18), (3.40), respectively.

With regard to Equations (3.48), (8.73), the elastic energy W_v, which is accumulated in the cell matrix *II*, is derived as

$$W_v = 8\,\Omega_v^{(12)}\left(\frac{u_{1c0}}{d}\right)^2. \quad (8.89)$$

With regard to Equation (8.77), the coefficient $\Omega_v^{(12)}$ is derived as

$$\Omega_v^{(12)} = \int_0^{\pi/2}\int_0^{\pi/2}\int_{R_p}^{r_c} \omega_v^{(12)}\, r^2\, dr\, d\varphi\, d\nu = \int_0^{\pi/2}\int_0^{\pi/2} \left\{\zeta_{1v}^2 U_{11v}^{(r)} + \zeta_{2v}^2 U_{22v}^{(r)} + \zeta_{1v}\zeta_{2v} U_{12v}^{(r)} + \frac{U_{11v}^{(\varphi\nu)}}{s_{44m}}\left[\left(\frac{\partial\zeta_{1v}}{\partial\varphi}\right)^2 + \left(\frac{\partial\zeta_{1v}}{\partial\nu}\right)^2\right]\right.$$

$$
+ \frac{U_{22v}^{(\varphi\nu)}}{s_{44m}} \left[\left(\frac{\partial \zeta_{2v}}{\partial \varphi} \right)^2 + \left(\frac{\partial \zeta_{2v}}{\partial \nu} \right)^2 \right]
+ \frac{U_{12v}^{(\varphi\nu)}}{s_{44m}} \left[\frac{\partial \zeta_{1v}}{\partial \varphi} \frac{\partial \zeta_{2v}}{\partial \varphi} + \frac{\partial \zeta_{1v}}{\partial \nu} \frac{\partial \zeta_{2v}}{\partial \nu} \right] \Bigg\} \, d\varphi \, d\nu, \quad (8.90)
$$

where $U_{ijv}^{(x)}$ $(i, j = 1,2,3;\ x = r, \varphi\nu)$ has the form

$$
U_{ijv}^{(x)} = \int_{R_v}^{r_c} u_{ijm}^{(x)} r^2 \, dr, \quad i, j = 1,2,3; \quad x = r, \ \varphi\nu. \quad (8.91)
$$

With regard to Equations (8.82), (8.91), $U_{ijv}^{(x)}$ $(i, j = 1,2;\ x = r,\ \varphi\nu)$ is given by Equations (8.83)–(8.88). In case of $U_{ijv}^{(x)}$, the radius R_p, which is included in Equations (8.83)–(8.88), is replaced by R_v.

With regard to Equations (5.13), (6.81), (8.78), (8.89), the displacement u_{1c0} along the axis x_i $(i = 1,2,3)$ on the cubic cell surface *I*, *II*, i.e., for $r = d/2$ (see Figure 2.1a), is given by Equation (6.89). The coefficients ζ_1, ζ_2, which are included in Equation (6.89), are derived as

$$
\zeta_1 = \frac{\Omega_{2m}^{(12)}}{2} + \frac{\Omega_{2p} R_p^3}{\xi_{3p}^2 (2\lambda_{1p} + 1)}, \quad (8.92)
$$

$$
\zeta_2 = \Omega_v^{(12)} - \left[\Omega_{1m}^{(12)} + \frac{\Omega_{1p} R_p^3}{\xi_{3p}^2 (2\lambda_{1p} + 1)} \right], \quad (8.93)
$$

where ξ_{3p}, Ω_{1p}, $\Omega_{1m}^{(12)}$, $\Omega_{2m}^{(12)}$, $\Omega_v^{(12)}$ are given by Equations (6.19), (8.79), (8.80), (8.90), respectively.

8.3. Integration Constants $C_{1q} \neq 0, C_{3q} \neq 0, C_{2q} = 0$

8.3.1. Cell Matrix around Particle

As presented above (see page 105), $u_{1rm} = u_{1rm}(r)$ is an increasing function of the variable r for $r < e^{1/3} \approx 1.39561$ m (see Equation (8.5)), and $u_{3rm} = u_{3rm}(r)$ is an increasing function of the variable r (see Equation (8.7)).

As presented in Section 5.2, the absolute values $|u'_{1m}|$, $\left|\varepsilon'_{ijm}\right|$, $\left|\sigma'_{ijm}\right|$ $(i, j = 1,2,3)$ are required to represent decreasing functions of $r \in \langle R_p, r_c \rangle$ (see

Equations (3.19)–(3.23), (3.36)–(3.39)). These decreasing functions exhibit maximum values on the particle-matrix boundary, i.e., for $r = R_p$. These decreasing courses are ensured by the integration constants $C_{1m} \neq 0$, $C_{3m} \neq 0$, which are determined by the mandatory boundary conditions (5.3), (5.4), where $C_{2m} = 0$. With regard to Equations (5.3), (5.4), (8.4), (8.5), (8.7), (8.15), we get

$$C_{1m} = \frac{1}{\xi_{1m}\zeta_m}\left[p_p\left(\frac{1}{2} + \ln r_c\right) + \frac{u'_{1c}\zeta_{1m}}{r_c}\right] = \frac{1}{\xi_{1m}\zeta_m}\left[p_p\left(\frac{1}{2} + \ln r_c\right) + \frac{u_{1c0}\zeta_{1m}}{d}\right], \tag{8.94}$$

$$C_{3m} = \frac{1}{\xi_{3m}\zeta_m}\left[p_p r_c\left(\frac{4}{3} - \ln r_c\right) + \frac{u'_{1c}\zeta_{2m}}{r_c}\right] = \frac{1}{\xi_{3m}\zeta_m}\left[p_p r_c\left(\frac{4}{3} - \ln r_c\right) + \frac{u_{1c0}\zeta_{2m}}{d}\right], \tag{8.95}$$

where $u'_{1c}/r_c = u_{1c0}/d$ (see Equations (3.15)–(3.18), (5.4)); and r_c, ξ_{im} $(i=1,3)$ are given by Equations (3.15)–(3.18), (7.8), (7.10), respectively. The radial stress p_p, which acts at the particle-matrix boundary, i.e., for $r = R_p$ (see Figure 2.1a), is determined in Section 8.3.3. The displacement u_{1c0} along the axis x_j $(j=1,2,3)$ on the cell surface *I*, *II* (see Equation (5.4)), i.e., for $r = d/2$ (see Figure 2.1), is determined in Section 8.3.4. The coefficients ζ_{im} $(i=1,2)$, ζ_m have the forms

$$\zeta_{1m} = \frac{r_c}{R_p}\left(c_{1m} - 2c_{2m}\ln R_p\right), \tag{8.96}$$

$$\zeta_{2m} = r_c\left[\frac{c_{1m} - 7c_{2m}}{3} - \left(c_{1m} - c_{2m}\right)\ln R_p\right], \tag{8.97}$$

$$\zeta_m = \frac{\zeta_{2m}}{r_c}\left(\frac{1}{2} + \ln r_c\right) - \zeta_{1m}\left(\frac{4}{3} - \ln r_c\right), \tag{8.98}$$

where c_{im} $(i=1,2)$ is given by Equation (3.40). With regard to Equations (7.4), (7.5), (7.7), (7.11)–(7.18), (8.94), (8.95), we get

$$u'_{1m} = -\frac{1}{\zeta_m}\left\{p_p\left[r\left(\frac{1}{2}+\ln r_c\right)\left(\frac{4}{3}-\ln r\right)-r_c\left(\frac{4}{3}-\ln r_c\right)\left(\frac{1}{2}+\ln r\right)\right]\right.$$
$$\left.+\frac{u_{1c0}}{d}\left[\zeta_{1m}\, r\left(\frac{4}{3}-\ln r\right)-\zeta_{2m}\left(\frac{1}{2}+\ln r\right)\right]\right\}, \qquad (8.99)$$

$$\varepsilon'_{11m} = -\frac{1}{\zeta_m}\left\{p_p\left[\left(\frac{1}{2}+\ln r_c\right)\left(\frac{1}{3}-\ln r\right)-\frac{r_c}{r}\left(\frac{4}{3}-\ln r_c\right)\right]\right.$$
$$\left.+\frac{u_{1c0}}{d}\left[\zeta_{1m}\left(\frac{1}{3}-\ln r\right)-\frac{\zeta_{2m}}{r}\right]\right\}, \qquad (8.100)$$

$$\varepsilon'_{22m} = \varepsilon'_{33m} =$$
$$-\frac{1}{\zeta_m}\left\{p_p\left[\left(\frac{1}{2}+\ln r_c\right)\left(\frac{4}{3}-\ln r\right)-\frac{r_c}{r}\left(\frac{4}{3}-\ln r_c\right)\left(\frac{1}{2}+\ln r\right)\right]\right.$$
$$\left.+\frac{u_{1c0}}{d}\left[\zeta_{1m}\left(\frac{4}{3}-\ln r\right)-\frac{\zeta_{2m}}{r}\left(\frac{1}{2}+\ln r\right)\right]\right\}, \qquad (8.101)$$

$$\varepsilon'_{12m} = -\frac{1}{\zeta_m}\left\{\left[\left(\frac{1}{2}+\ln r_c\right)\left(\frac{4}{3}-\ln r\right)-\frac{r_c}{r}\left(\frac{4}{3}-\ln r_c\right)\left(\frac{1}{2}+\ln r\right)\right]\right.$$
$$\times\left(\frac{\partial p_p}{\partial\varphi}-\frac{p_p}{\zeta_m}\frac{\partial\zeta_m}{\partial\varphi}\right)$$
$$+p_p\left[\frac{1}{r_c}\left(\frac{4}{3}-\ln r\right)-\frac{1}{r}\left(\frac{1}{3}\quad\ln r_c\right)\left(\frac{1}{2}+\ln r\right)\right]\frac{\partial r_c}{\partial\varphi}$$
$$+\frac{u_{1c0}}{d}\left[\zeta_{1m}\left(\frac{4}{3}-\ln r\right)\left(\frac{\partial\zeta_{1m}}{\partial\varphi}-\frac{\zeta_{1m}}{\zeta_m}\frac{\partial\zeta_m}{\partial\varphi}\right)\right.$$
$$\left.\left.-\frac{\zeta_{2m}}{r}\left(\frac{1}{2}+\ln r\right)\left(\frac{\partial\zeta_{2m}}{\partial\varphi}-\frac{\zeta_{2m}}{\zeta_m}\frac{\partial\zeta_m}{\partial\varphi}\right)\right]\right\},$$
$$(8.102)$$

$$\varepsilon'_{13m} = -\frac{1}{\zeta_m}\left\{\left[\left(\frac{1}{2}+\ln r_c\right)\left(\frac{4}{3}-\ln r\right)-\frac{r_c}{r}\left(\frac{4}{3}-\ln r_c\right)\left(\frac{1}{2}+\ln r\right)\right]\right.$$

$$\times\left(\frac{\partial p_p}{\partial \nu}-\frac{p_p}{\zeta_m}\frac{\partial \zeta_m}{\partial \nu}\right)$$

$$+\,p_p\left[\frac{1}{r_c}\left(\frac{4}{3}-\ln r\right)-\frac{1}{r}\left(\frac{1}{3}-\ln r_c\right)\left(\frac{1}{2}+\ln r\right)\right]\frac{\partial r_c}{\partial \nu}$$

$$+\,\frac{u_{1c0}}{d}\left[\zeta_{1m}\left(\frac{4}{3}-\ln r\right)\left(\frac{\partial \zeta_{1m}}{\partial \nu}-\frac{\zeta_{1m}}{\zeta_m}\frac{\partial \zeta_m}{\partial \nu}\right)\right.$$

$$\left.\left.-\,\frac{\zeta_{2m}}{r}\left(\frac{1}{2}+\ln r\right)\left(\frac{\partial \zeta_{2m}}{\partial \nu}-\frac{\zeta_{2m}}{\zeta_m}\frac{\partial \zeta_m}{\partial \nu}\right)\right]\right\},\qquad (8.103)$$

$$\sigma'_{11m}=\frac{1}{\zeta_m}\left(p_p\left\{\left(\frac{1}{2}+\ln r_c\right)\left[\frac{c_{1m}-7c_{2m}}{3}-(c_{1m}-c_{2m})\ln r\right]\right.\right.$$

$$\left.+\,\frac{r_c}{r}\left(\frac{4}{3}-\ln r_c\right)(c_{1m}-2c_{2m}\ln r)\right\}$$

$$+\,\frac{u_{1c0}}{d}\left\{\zeta_{1m}\left[\frac{c_{1m}-7c_{2m}}{3}-(c_{1m}-c_{2m})\ln r\right]\right.$$

$$\left.\left.+\,\frac{\zeta_{2m}}{r}(c_{1m}-2c_{2m}\ln r)\right\}\right),\qquad (8.104)$$

$$\sigma'_{22m}=\sigma'_{33m}=\frac{1}{\zeta_m}\left(p_p\left\{\left(\frac{1}{2}+\ln r_c\right)\left[\frac{4c_{1m}-c_{2m}}{3}-(c_{1m}-c_{2m})\ln r\right]\right.\right.$$

$$\left.+\,\frac{r_c}{r}\left(\frac{4}{3}-\ln r_c\right)\left(\frac{c_{1m}-c_{2m}}{2}+c_{3m}\ln r\right)\right\}$$

$$+\,\frac{u_{1c0}}{d}\left\{\zeta_{1m}\left[\frac{4c_{1m}-c_{2m}}{3}-(c_{1m}-c_{2m})\ln r\right]\right.$$

$$\left.\left.+\,\frac{\zeta_{2m}}{r}\left(\frac{c_{1m}-c_{2m}}{2}+c_{3m}\ln r\right)\right\}\right),\qquad (8.105)$$

$$\sigma'_{12m}=-\frac{1}{s_{44m}\zeta_m}\left\{\left[\left(\frac{1}{2}+\ln r_c\right)\left(\frac{4}{3}-\ln r\right)-\frac{r_c}{r}\left(\frac{4}{3}-\ln r_c\right)\left(\frac{1}{2}+\ln r\right)\right]\right.$$

$$
\times \left(\frac{\partial p_p}{\partial \varphi} - \frac{p_p}{\zeta_m} \frac{\partial \zeta_m}{\partial \varphi} \right)
+ p_p \left[\frac{1}{r_c} \left(\frac{4}{3} - \ln r \right) - \frac{1}{r} \left(\frac{1}{3} - \ln r_c \right) \left(\frac{1}{2} + \ln r \right) \right] \frac{\partial r_c}{\partial \varphi}
+ \frac{u_{1c0}}{d} \left[\zeta_{1m} \left(\frac{4}{3} - \ln r \right) \left(\frac{\partial \zeta_{1m}}{\partial \varphi} - \frac{\zeta_{1m}}{\zeta_m} \frac{\partial \zeta_m}{\partial \varphi} \right)
- \frac{\zeta_{2m}}{r} \left(\frac{1}{2} + \ln r \right) \left(\frac{\partial \zeta_{2m}}{\partial \varphi} - \frac{\zeta_{2m}}{\zeta_m} \frac{\partial \zeta_m}{\partial \varphi} \right) \right] \Bigg\},
\tag{8.106}
$$

$$
\sigma'_{13m} = - \frac{1}{s_{44m} \zeta_m} \Bigg\{ \left[\left(\frac{1}{2} + \ln r_c \right) \left(\frac{4}{3} - \ln r \right) - \frac{r_c}{r} \left(\frac{4}{3} - \ln r_c \right) \left(\frac{1}{2} + \ln r \right) \right]
\times \left(\frac{\partial p_p}{\partial \nu} - \frac{p_p}{\zeta_m} \frac{\partial \zeta_m}{\partial \nu} \right)
+ p_p \left[\frac{1}{r_c} \left(\frac{4}{3} - \ln r \right) - \frac{1}{r} \left(\frac{1}{3} - \ln r_c \right) \left(\frac{1}{2} + \ln r \right) \right] \frac{\partial r_c}{\partial \nu}
+ \frac{u_{1c0}}{d} \left[\zeta_{1m} \left(\frac{4}{3} - \ln r \right) \left(\frac{\partial \zeta_{1m}}{\partial \nu} - \frac{\zeta_{1m}}{\zeta_m} \frac{\partial \zeta_m}{\partial \nu} \right)
- \frac{\zeta_{2m}}{r} \left(\frac{1}{2} + \ln r \right) \left(\frac{\partial \zeta_{2m}}{\partial \nu} - \frac{\zeta_{2m}}{\zeta_m} \frac{\partial \zeta_m}{\partial \nu} \right) \right] \Bigg\},
\tag{8.107}
$$

where s_{44m} is given by Equation (3.35).

8.3.2. Cell Matrix around Void

As presented in Section 5.3, the absolute values $|u'_{1v}|$, $\left|\varepsilon'_{ijv}\right|$, $\left|\sigma'_{ijv}\right|$ $(i, j = 1,2,3)$ are required to represent increasing functions of $r \in \langle R_v, r_c \rangle$ (see Equations (3.19)–(3.23), (3.36)–(3.39)). These increasing functions exhibit maximum values on the cell surface, i.e., for $r = r_c$ (see Equations (3.15)–(3.18)). These increasing courses are ensured by the integration constants $C_{1v} \neq 0$, $C_{3v} \neq 0$, which are determined by the mandatory boundary conditions (5.5), (5.6), where $C_{2v} = 0$. With regard to Equations (5.5), (5.6), (8.4), (8.5), (8.7), (8.15), we get

$$C_{1v} = -\frac{u'_{1c}\chi_{1v}}{\xi_{1v}\chi_v r_c} = -\frac{u_{1c0}\chi_{1v}}{d\xi_{1v}\chi_v}, \tag{8.108}$$

$$C_{3v} = -\frac{u'_{1c}\chi_{2v}}{\xi_{3v}\chi_v r_c} = -\frac{u_{1c0}\chi_{2v}}{d\xi_{3v}\chi_v}, \tag{8.109}$$

where $u'_{1c}/r_c = u_{1c0}/d$ (see Equations (3.15)–(3.18), (5.4)); and r_c, ξ_{iv} $(i = 1,3)$ are given by Equations (3.15)–(3.18), (8.8), (8.10), respectively. The displacement u_{1c0} along the axis x_j $(j = 1,2,3)$ on the cell surface *I*, *II* (see Equation (5.4)), i.e., for $r = d/2$ (see Figure 2.1), is determined in Section 8.2.4. The coefficients χ_{iv} $(i = 1,2)$, χ_v have the forms

$$\chi_{1v} = \frac{r_c}{R_v}\left(c_{1m} - 2c_{2m}\ln R_v\right), \tag{8.110}$$

$$\chi_{2v} = r_c\left[\frac{c_{1m} - 7c_{2m}}{3} - (c_{1m} - c_{2m})\ln R_v\right], \tag{8.111}$$

$$\chi_v = \frac{\chi_{2v}}{r_c}\left(\frac{1}{2} + \ln r_c\right) - \chi_{1v}\left(\frac{4}{3} - \ln r_c\right), \tag{8.112}$$

where c_{im} $(i = 1,2)$ is given by Equation (3.40). With regard to Equations (7.4), (7.5), (7.7), (7.11)–(7.18), (8.94), (8.95), we get

$$u'_{1v} = \frac{u_{1c0}}{d\chi_v}\left[\chi_{1v} r\left(\frac{4}{3} - \ln r\right) - \chi_{2v}\left(\frac{1}{2} + \ln r\right)\right], \tag{8.113}$$

$$\varepsilon'_{11v} = \frac{u_{1c0}}{d\chi_v}\left[\chi_{1v}\left(\frac{1}{3} - \ln r\right) - \frac{\chi_{2v}}{r}\right], \tag{8.114}$$

$$\varepsilon'_{22v} = \varepsilon'_{33v} = \frac{u_{1c0}}{d\chi_v}\left[\chi_{1v}\left(\frac{4}{3} - \ln r\right) - \frac{\chi_{2v}}{r}\left(\frac{1}{2} + \ln r\right)\right], \tag{8.115}$$

$$\varepsilon'_{12v} = \frac{u_{1c0}}{d\chi_v}\left[\chi_{1v}\left(\frac{4}{3} - \ln r\right)\left(\frac{\partial\chi_{1v}}{\partial\varphi} - \frac{\chi_{1v}}{\chi_v}\frac{\partial\chi_v}{\partial\varphi}\right) - \frac{\chi_{2v}}{r}\left(\frac{1}{2} + \ln r\right)\left(\frac{\partial\chi_{2v}}{\partial\varphi} - \frac{\chi_{2v}}{\chi_v}\frac{\partial\chi_v}{\partial\varphi}\right)\right], \tag{8.116}$$

$$\varepsilon'_{13v} = \frac{u_{1c0}}{d\chi_v}\left[\chi_{1v}\left(\frac{4}{3} - \ln r\right)\left(\frac{\partial \chi_{1v}}{\partial \nu} - \frac{\chi_{1v}}{\chi_v}\frac{\partial \chi_v}{\partial \nu}\right)\right.$$
$$\left. - \frac{\chi_{2v}}{r}\left(\frac{1}{2} + \ln r\right)\left(\frac{\partial \chi_{2v}}{\partial \nu} - \frac{\chi_{2v}}{\chi_v}\frac{\partial \chi_v}{\partial \nu}\right)\right], \quad (8.117)$$

$$\sigma'_{11v} = -\frac{u_{1c0}}{d\chi_v}\left\{\chi_{1v}\left[\frac{c_{1m} - 7c_{2m}}{3} - (c_{1m} - c_{2m})\ln r\right] + \frac{\chi_{2v}}{r}(c_{1m} - 2c_{2m}\ln r)\right\}, \quad (8.118)$$

$$\sigma'_{22v} = \sigma'_{33v} = -\frac{u_{1c0}}{d\chi_v}\left\{\chi_{1v}\left[\frac{4c_{1m} - c_{2m}}{3} - (c_{1m} - c_{2m})\ln r\right]\right.$$
$$\left. + \frac{\chi_{2v}}{r}\left(\frac{c_{1m} - c_{2m}}{2} + c_{3m}\ln r\right)\right\}, \quad (8.119)$$

$$\sigma'_{12v} = \frac{u_{1c0}}{d s_{44m}\chi_v}\left[\chi_{1v}\left(\frac{4}{3} - \ln r\right)\left(\frac{\partial \chi_{1v}}{\partial \varphi} - \frac{\chi_{1v}}{\chi_v}\frac{\partial \chi_v}{\partial \varphi}\right)\right.$$
$$\left. - \frac{\chi_{2v}}{r}\left(\frac{1}{2} + \ln r\right)\left(\frac{\partial \chi_{2v}}{\partial \varphi} - \frac{\chi_{2v}}{\chi_v}\frac{\partial \chi_v}{\partial \varphi}\right)\right], \quad (8.120)$$

$$\sigma'_{13v} = \frac{u_{1c0}}{d s_{44m}\chi_v}\left[\chi_{1v}\left(\frac{4}{3} - \ln r\right)\left(\frac{\partial \chi_{1v}}{\partial \nu} - \frac{\chi_{1v}}{\chi_v}\frac{\partial \chi_v}{\partial \nu}\right)\right.$$
$$\left. - \frac{\chi_{2v}}{r}\left(\frac{1}{2} + \ln r\right)\left(\frac{\partial \chi_{2v}}{\partial \nu} - \frac{\chi_{2v}}{\chi_v}\frac{\partial \chi_v}{\partial \nu}\right)\right], \quad (8.121)$$

where s_{44m} is given by Equation (3.35).

8.3.3. Radial Stress p_p

With regard to Equations (5.12), (6.25), (8.101), the radial stress p_p, which acts at the particle-matrix boundary, i.e., for $r = R_p$ (see Figure 2.1a), is given by Equations (8.60)–(8.63). The coefficients ζ_{3m}, ζ_{4m}, which are included in Equations (8.60), (8.61), have the forms

$$\zeta_{3m} = \frac{1}{\zeta_m} \left[\left(\frac{1}{2} + \ln r_c \right) \left(\frac{4}{3} - \ln R_p \right) - \frac{r_c}{R_p} \left(\frac{4}{3} - \ln r_c \right) \left(\frac{1}{2} + \ln R_p \right) \right], \tag{8.122}$$

$$\zeta_{4m} = \frac{1}{\zeta_m} \left[\zeta_{1m} \left(\frac{4}{3} - \ln R_p \right) - \frac{\zeta_{2m}}{R_p} \left(\frac{1}{2} + \ln R_p \right) \right], \tag{8.123}$$

where r_c, ζ_{1m}, ζ_{2m}, ζ_m are given by Equations (3.15)–(3.18), (8.96)–(8.98), respectively.

8.3.4. Displacement u_{1c0}

With regard to Equations (8.60), (8.94), (8.95), (8.108), (8.109), the integration constant C_{iq} ($i = 1,3$; $q = m,v$) has the form

$$C_{im} = \frac{1}{\xi_{im}} \left[\frac{u_{1c0} \zeta_{i1m}}{d} + \zeta_{i2m} \left(\beta_m - \beta_p \right) \right], \quad i = 1,3, \tag{8.124}$$

$$C_{iv} = \frac{u_{1c0} \zeta_{iv}}{d \xi_{iv}}, \quad i = 1,3, \tag{8.125}$$

where ξ_{im}, ξ_{iv} ($i = 1,3$) are given by Equations (7.8), (7.10), and β_q ($q = p,m$) is determined in Section 4.2. The coefficients ζ_{i1m}, ζ_{i2m}, ζ_{iv} ($i = 1,3$) are derived as

$$\zeta_{11m} = \frac{1}{\zeta_m} \left[\zeta_{1m} + \zeta_{pm} \zeta_{4m} \left(\frac{1}{2} + \ln r_c \right) \right], \tag{8.126}$$

$$\zeta_{31m} = \frac{1}{\zeta_m} \left[\zeta_{2m} + \zeta_{pm} \zeta_{4m} \left(\frac{4}{3} - \ln r_c \right) \right], \tag{8.127}$$

$$\zeta_{12m} = \frac{\zeta_{pm}}{\zeta_m} \left(\frac{1}{2} + \ln r_c \right), \tag{8.128}$$

$$\zeta_{32m} = \frac{\zeta_{pm}}{\zeta_m} \left(\frac{4}{3} - \ln r_c \right), \tag{8.129}$$

$$\zeta_{1v} = -\frac{\chi_{1v}}{\chi_v}, \tag{8.130}$$

$$\zeta_{3v} = -\frac{\chi_{2v}}{\chi_v}, \tag{8.131}$$

where r_c, ζ_{im}, ζ_m, χ_{iv} $(i=1,2,4)$, χ_v and ζ_{pm}, ζ_{4m} are given by Equations (3.15)–(3.18), (8.96)–(8.98), (8.123), (8.110)–(8.112) and (8.60), (8.122), respectively. The coefficients ζ_{3m}, ζ_{4m}, which are included in ζ_{pm} (see Equation (7.60)), are given by Equations (7.122), (7.123), respectively.

With regard to Equations (8.19), (8.20), (8.22), (8.25), (8.26), (8.28), (8.31), (8.124), (8.125), the elastic energy density w_q $(q=m,v)$ is given by Equations (8.72), (8.73). The coefficients $\omega_{im}^{(12)}$ $(i=1,2,3)$, $\omega_v^{(12)}$, which are included in Equations (8.72), (8.73), are replaced by $\omega_{im}^{(13)}$ $(i=1,2,3)$, $\omega_v^{(13)}$, respectively. The coefficients $\omega_{im}^{(13)}$ $(i=1,2,3)$, $\omega_v^{(13)}$ are derived as

$$\begin{aligned}\omega_{1m}^{(13)} &= \zeta_{11m}^2 u_{11m}^{(r)} + \zeta_{31m}^2 u_{33m}^{(r)} + \zeta_{11m}\zeta_{31m} u_{13m}^{(r)} \\ &+ \frac{u_{11m}^{(\varphi\nu)}}{s_{44m}}\left[\left(\frac{\partial\zeta_{11m}}{\partial\varphi}\right)^2 + \left(\frac{\partial\zeta_{11m}}{\partial\nu}\right)^2\right] \\ &+ \frac{u_{33m}^{(\varphi\nu)}}{s_{44m}}\left[\left(\frac{\partial\zeta_{31m}}{\partial\varphi}\right)^2 + \left(\frac{\partial\zeta_{31m}}{\partial\nu}\right)^2\right] \\ &+ \frac{u_{13m}^{(\varphi\nu)}}{s_{44m}}\left[\frac{\partial\zeta_{11m}}{\partial\varphi}\frac{\partial\zeta_{31m}}{\partial\varphi} + \frac{\partial\zeta_{11m}}{\partial\nu}\frac{\partial\zeta_{31m}}{\partial\nu}\right],\end{aligned} \tag{8.132}$$

$$\begin{aligned}\omega_{2m}^{(13)} &= 2\left[\zeta_{11m}\zeta_{12m} u_{11m}^{(r)} + \zeta_{31m}\zeta_{32m} u_{33m}^{(r)}\right] + \left(\zeta_{11m}\zeta_{32m} + \zeta_{12m}\zeta_{31m}\right) u_{13m}^{(r)} \\ &+ \frac{2u_{11m}^{(\varphi\nu)}}{s_{44m}}\left(\frac{\partial\zeta_{11m}}{\partial\varphi}\frac{\partial\zeta_{12m}}{\partial\varphi} + \frac{\partial\zeta_{11m}}{\partial\nu}\frac{\partial\zeta_{12m}}{\partial\nu}\right) \\ &+ \frac{2u_{33m}^{(\varphi\nu)}}{s_{44m}}\left(\frac{\partial\zeta_{31m}}{\partial\varphi}\frac{\partial\zeta_{32m}}{\partial\varphi} + \frac{\partial\zeta_{31m}}{\partial\nu}\frac{\partial\zeta_{32m}}{\partial\nu}\right) \\ &+ \frac{u_{13m}^{(\varphi\nu)}}{s_{44m}}\left(\frac{\partial\zeta_{11m}}{\partial\varphi}\frac{\partial\zeta_{32m}}{\partial\varphi} + \frac{\partial\zeta_{12m}}{\partial\varphi}\frac{\partial\zeta_{31m}}{\partial\varphi}\right. \\ &\qquad \left. + \frac{\partial\zeta_{11m}}{\partial\nu}\frac{\partial\zeta_{32m}}{\partial\nu} + \frac{\partial\zeta_{12m}}{\partial\nu}\frac{\partial\zeta_{31m}}{\partial\nu}\right),\end{aligned} \tag{8.133}$$

$$\omega_{3m}^{(13)} = \zeta_{12m}^2 u_{11m}^{(r)} + \zeta_{32m}^2 u_{33m}^{(r)} + \zeta_{12m}\zeta_{32m} u_{13m}^{(r)} + \frac{u_{11m}^{(\varphi\nu)}}{s_{44m}} \left[\left(\frac{\partial \zeta_{12m}}{\partial \varphi} \right)^2 + \left(\frac{\partial \zeta_{12m}}{\partial \nu} \right)^2 \right] + \frac{u_{33m}^{(\varphi\nu)}}{s_{44m}} \left[\left(\frac{\partial \zeta_{32m}}{\partial \varphi} \right)^2 + \left(\frac{\partial \zeta_{32m}}{\partial \nu} \right)^2 \right] + \frac{u_{13m}^{(\varphi\nu)}}{s_{44m}} \left[\frac{\partial \zeta_{12m}}{\partial \varphi}\frac{\partial \zeta_{32m}}{\partial \varphi} + \frac{\partial \zeta_{12m}}{\partial \nu}\frac{\partial \zeta_{32m}}{\partial \nu} \right], \tag{8.134}$$

$$\omega_{\nu}^{(13)} = \zeta_{1\nu}^2 u_{11m}^{(r)} + \zeta_{3\nu}^2 u_{33m}^{(r)} + \zeta_{1\nu}\zeta_{3\nu} u_{13m}^{(r)} + \frac{u_{11m}^{(\varphi\nu)}}{s_{44m}} \left[\left(\frac{\partial \zeta_{1\nu}}{\partial \varphi} \right)^2 + \left(\frac{\partial \zeta_{1\nu}}{\partial \nu} \right)^2 \right] + \frac{u_{33m}^{(\varphi\nu)}}{s_{44m}} \left[\left(\frac{\partial \zeta_{3\nu}}{\partial \varphi} \right)^2 + \left(\frac{\partial \zeta_{3\nu}}{\partial \nu} \right)^2 \right] + \frac{u_{13m}^{(\varphi\nu)}}{s_{44m}} \left[\frac{\partial \zeta_{1\nu}}{\partial \varphi}\frac{\partial \zeta_{3\nu}}{\partial \varphi} + \frac{\partial \zeta_{1\nu}}{\partial \nu}\frac{\partial \zeta_{3\nu}}{\partial \nu} \right], \tag{8.135}$$

where s_{44m} is given by Equation (3.35).

With regard to Equations (3.47), (8.72), the elastic energy W_m, which is accumulated in the cell matrix I, is given by Equation (8.78). With regard to Equations (8.72), (8.132)–(8.134), the coefficient $\Omega_{im}^{(12)}$ $(i = 1,2,3)$, which is included in Equation (8.78), is replaced by $\Omega_{im}^{(13)}$ $(i = 1,2,3)$. The coefficient $\Omega_{im}^{(13)}$ $(i = 1,2,3)$ has the form

$$\Omega_{1m}^{(13)} = \int_0^{\pi/2}\int_0^{\pi/2}\int_{R_p}^{r_c} \omega_{1m}^{(13)} r^2 \, dr \, d\varphi \, d\nu = \int_0^{\pi/2}\int_0^{\pi/2} \left\{ \zeta_{11m}^2 U_{11m}^{(r)} + \zeta_{31m}^2 U_{33m}^{(r)} + \zeta_{11m}\zeta_{31m} U_{13m}^{(r)} \right.$$

$$+\frac{U_{11m}^{(\varphi\nu)}}{s_{44m}}\left[\left(\frac{\partial\zeta_{11m}}{\partial\varphi}\right)^2+\left(\frac{\partial\zeta_{11m}}{\partial\nu}\right)^2\right]$$
$$+\frac{U_{33m}^{(\varphi\nu)}}{s_{44m}}\left[\left(\frac{\partial\zeta_{31m}}{\partial\varphi}\right)^2+\left(\frac{\partial\zeta_{31m}}{\partial\nu}\right)^2\right]$$
$$\left.+\frac{U_{13m}^{(\varphi\nu)}}{s_{44m}}\left[\frac{\partial\zeta_{11m}}{\partial\varphi}\frac{\partial\zeta_{31m}}{\partial\varphi}+\frac{\partial\zeta_{11m}}{\partial\nu}\frac{\partial\zeta_{31m}}{\partial\nu}\right]\right\}d\varphi\,d\nu, \quad (8.136)$$

$$\Omega_{2m}^{(13)}=\int_0^{\pi/2}\int_0^{\pi/2}\int_{R_p}^{r_c}\omega_{2m}^{(13)}\,r^2\,dr\,d\varphi\,d\nu$$
$$=\int_0^{\pi/2}\int_0^{\pi/2}\left\{2\left[\zeta_{11m}\zeta_{12m}U_{11m}^{(r)}+\zeta_{31m}\zeta_{32m}U_{33m}^{(r)}\right]\right.$$
$$+\left(\zeta_{11m}\zeta_{32m}+\zeta_{12m}\zeta_{31m}\right)U_{13m}^{(r)}$$
$$+\frac{2U_{11m}^{(\varphi\nu)}}{s_{44m}}\left(\frac{\partial\zeta_{11m}}{\partial\varphi}\frac{\partial\zeta_{12m}}{\partial\varphi}+\frac{\partial\zeta_{11m}}{\partial\nu}\frac{\partial\zeta_{12m}}{\partial\nu}\right)$$
$$+\frac{2U_{33m}^{(\varphi\nu)}}{s_{44m}}\left(\frac{\partial\zeta_{31m}}{\partial\varphi}\frac{\partial\zeta_{32m}}{\partial\varphi}+\frac{\partial\zeta_{31m}}{\partial\nu}\frac{\partial\zeta_{32m}}{\partial\nu}\right)$$
$$+\frac{U_{13m}^{(\varphi\nu)}}{s_{44m}}\left(\frac{\partial\zeta_{11m}}{\partial\varphi}\frac{\partial\zeta_{32m}}{\partial\varphi}+\frac{\partial\zeta_{12m}}{\partial\varphi}\frac{\partial\zeta_{31m}}{\partial\varphi}\right)$$
$$\left.+\frac{U_{13m}^{(\varphi\nu)}}{s_{44m}}\left(\frac{\partial\zeta_{11m}}{\partial\nu}\frac{\partial\zeta_{32m}}{\partial\nu}+\frac{\partial\zeta_{12m}}{\partial\nu}\frac{\partial\zeta_{31m}}{\partial\nu}\right)\right\}d\varphi\,d\nu, \quad (8.137)$$

$$\Omega_{3m}^{(13)}=\int_0^{\pi/2}\int_0^{\pi/2}\int_{R_p}^{r_c}\omega_{3m}^{(13)}\,r^2\,dr\,d\varphi\,d\nu$$
$$=\int_0^{\pi/2}\int_0^{\pi/2}\left\{\zeta_{12m}^2U_{11m}^{(r)}+\zeta_{32m}^2U_{33m}^{(r)}+\zeta_{12m}\zeta_{32m}U_{13m}^{(r)}\right.$$

$$+\frac{U_{11m}^{(\varphi\nu)}}{s_{44m}}\left[\left(\frac{\partial\zeta_{12m}}{\partial\varphi}\right)^2+\left(\frac{\partial\zeta_{12m}}{\partial\nu}\right)^2\right]$$
$$+\frac{U_{33m}^{(\varphi\nu)}}{s_{44m}}\left[\left(\frac{\partial\zeta_{32m}}{\partial\varphi}\right)^2+\left(\frac{\partial\zeta_{32m}}{\partial\nu}\right)^2\right]$$
$$\left.+\frac{U_{13m}^{(\varphi\nu)}}{s_{44m}}\left[\frac{\partial\zeta_{12m}}{\partial\varphi}\frac{\partial\zeta_{32m}}{\partial\varphi}+\frac{\partial\zeta_{12m}}{\partial\nu}\frac{\partial\zeta_{32m}}{\partial\nu}\right]\right\} d\varphi\, d\nu, \quad (8.138)$$

where $U_{ijm}^{(x)}$ $(i,j=1,3;\ x=r,\varphi\nu)$, $U_{11m}^{(r)}$, $U_{11m}^{(\varphi\nu)}$ are given by Equations (7.82), (7.83), (7.86), respectively. With regard to Equations (8.22), (8.25), (8.28), (8.31), (8.82), we get

$$U_{13m}^{(r)}=\frac{3c_{1m}-c_{2m}}{2}\left[r_c^2\left(\ln r_c-\frac{1}{2}\right)-R_p^2\left(\ln R_p-\frac{1}{2}\right)\right]$$
$$-\frac{4c_{1m}-c_{2m}}{6}\left(r_c^2-R_p^2\right), \quad (8.139)$$

$$U_{33m}^{(r)}=c_{1m}\left[r_c\left(\ln^2 r_c-2\ln r_c+2\right)-R_p\left(\ln^2 R_p-2\ln R_p+2\right)\right]$$
$$-c_{1m}\left[r_c\left(\ln r_c-1\right)-R_p\left(\ln R_p-1\right)\right]+\frac{c_{2m}-2c_{1m}}{4}\left(r_c-R_p\right), \quad (8.140)$$

$$U_{13m}^{(\varphi\nu)}=\frac{2\left(r_c^2-R_p^2\right)}{3}+\frac{5}{6}\left[r_c^2\left(\ln r_c-\frac{1}{2}\right)-R_p^2\left(\ln R_p-\frac{1}{2}\right)\right]$$
$$-r_c^2\left(\ln^2 r_c-\ln r_c+\frac{1}{2}\right)+R_p^2\left(\ln^2 R_p-\ln R_p+\frac{1}{2}\right), \quad (8.141)$$

$$U_{33m}^{(\varphi\nu)}=r_c\ln r_c\left(\ln r_c-1\right)-R_p\ln R_p\left(\ln R_p-1\right)+\frac{5\left(r_c-R_p\right)}{4}, \quad (8.142)$$

where r_c, c_{im} $(i=1,2)$ are given by Equations (3.15)–(3.18), (3.40), respectively.

With regard to Equations (3.48), (8.73), the elastic energy W_v, which is accumulated in the cell matrix *II*, is given by Equation (8.89). With regard to Equations (8.73), (8.135), the coefficient $\Omega_v^{(12)}$, which is included in Equation (8.89), is replaced by $\Omega_v^{(13)}$. The coefficient $\Omega_v^{(13)}$ has the form

$$\Omega_{v}^{(13)} = \int_{0}^{\pi/2}\int_{0}^{\pi/2}\int_{R_p}^{r_c} \omega_{v}^{(13)} r^2 \, dr \, d\varphi \, d\nu$$
$$= \int_{0}^{\pi/2}\int_{0}^{\pi/2} \left\{ \zeta_{1\nu}^2 U_{11\nu}^{(r)} + \zeta_{3\nu}^2 U_{33\nu}^{(r)} + \zeta_{1\nu}\zeta_{3\nu} U_{13\nu}^{(r)} \right.$$
$$+ \frac{U_{11\nu}^{(\varphi\nu)}}{s_{44m}} \left[\left(\frac{\partial \zeta_{1\nu}}{\partial \varphi}\right)^2 + \left(\frac{\partial \zeta_{1\nu}}{\partial \nu}\right)^2 \right]$$
$$+ \frac{U_{33\nu}^{(\varphi\nu)}}{s_{44m}} \left[\left(\frac{\partial \zeta_{3\nu}}{\partial \varphi}\right)^2 + \left(\frac{\partial \zeta_{3\nu}}{\partial \nu}\right)^2 \right]$$
$$\left. + \frac{U_{13\nu}^{(\varphi\nu)}}{s_{44m}} \left[\frac{\partial \zeta_{1\nu}}{\partial \varphi}\frac{\partial \zeta_{3\nu}}{\partial \varphi} + \frac{\partial \zeta_{1\nu}}{\partial \nu}\frac{\partial \zeta_{3\nu}}{\partial \nu} \right] \right\} d\varphi \, d\nu. \quad (8.143)$$

With regard to Equations (8.82), (8.91), $U_{ij\nu}^{(x)}$ $(i, j = 1,3;\ x = r,\ \varphi\nu)$ is given by Equations (8.139)–(8.142). In case of $U_{ij\nu}^{(x)}$, the radius R_p, which is included in Equations (8.139)–(8.142), is replaced by R_ν.

With regard to Equations (5.13), (6.81), (8.78), (8.89), the displacement u_{1c0} along the axis x_i $(i = 1,2,3)$ on the cubic cell surface *I*, *II*, i.e., for $r = d/2$ (see Figure 2.1a), is given by Equations (6.89), (8.92), (8.93).

The coefficients $\Omega_{1m}^{(12)}$, $\Omega_{2m}^{(12)}$, $\Omega_{v}^{(12)}$, which are included in Equations (8.92), (8.93), are replaced by $\Omega_{1m}^{(13)}$, $\Omega_{2m}^{(13)}$, $\Omega_{v}^{(13)}$ (see Equations (8.136), (8.137), (8.143)), respectively.

8.4. Integration Constants $C_{2q} \neq 0$, $C_{3q} \neq 0$, $C_{1q} = 0$

8.4.1. Cell Matrix around Particle

As presented above (see page 55), $u_{2rm} = u_{2rm}(r)$ is a decreasing function of the variable r due to $c_{3m} < 0$, and u_{3rm} is an increasing function of the variable r (see Equations (3.40), (8.6), (8.7)).

As presented in Section 5.2, the absolute values $|u'_{1m}|$, $\left|\varepsilon'_{ijm}\right|$, $\left|\sigma'_{ijm}\right|$ $(i, j = 1,2,3)$ are required to represent decreasing functions of $r \in \langle R_p, r_c \rangle$ (see

Equations (3.19)–(3.23), (3.36)–(3.39)). These decreasing functions exhibit maximum values on the particle-matrix boundary, i.e., for $r = R_p$. These decreasing courses are ensured by the integration constants $C_{2m} \neq 0$, $C_{3m} \neq 0$, which are determined by the mandatory boundary conditions (5.3), (5.4), where $C_{1m} = 0$. With regard to Equations (5.3), (5.4), (8.4), (8.6), (8.7), (8.15), we get

$$C_{2m} = -\frac{1}{\xi_{2m}\zeta_m}\left[p_p\left(\frac{1}{2}+\ln r_c\right)+\frac{u'_{1c}\zeta_{1m}}{r_c}\right] = -\frac{1}{\xi_{2m}\zeta_m}\left[p_p\left(\frac{1}{2}+\ln r_c\right)+\frac{u_{1c0}\zeta_{1m}}{d}\right], \quad (8.144)$$

$$C_{3m} = \frac{1}{\xi_{3m}\zeta_m}\left(p_p r_c^{c_{3m}}+\frac{u'_{1c}\zeta_{2m}}{r_c}\right) = \frac{1}{\xi_{3m}\zeta_m}\left(p_p r_c^{c_{3m}}+\frac{u_{1c0}\zeta_{2m}}{d}\right), \quad (8.145)$$

where $u'_{1c}/r_c = u_{1c0}/d$ (see Equations (3.15)–(3.18), (5.4)); and r_c, ξ_{im} $(i=2,3)$ are given by Equations (3.15)–(3.18), (8.9), (8.10), respectively. The radial stress p_p, which acts at the particle-matrix boundary, i.e., for $r = R_p$ (see Figure 2.1a), is determined in Section 8.4.3. The displacement u_{1c0} along the axis x_j $(j=1,2,3)$ on the cell surface *I*, *II* (see Equation (5.4)), i.e., for $r = d/2$ (see Figure 2.1), is determined in Section 8.4.4. The coefficients ζ_{im} $(i=1,2)$, ζ_m have the forms

$$\zeta_{1m} = \frac{r_c}{R_p}\left(c_{1m} - 2c_{2m}\ln R_p\right), \quad (8.146)$$

$$\zeta_{2m} = r_c\left[c_{3m}\left(c_{1m}+c_{2m}\right) - 2c_{2m}\right]R_p^{c_{3m}-1}, \quad (8.147)$$

$$\zeta_m = \frac{\zeta_{2m}}{r_c}\left(\frac{1}{2}+\ln r_c\right) - \zeta_{1m}r_c^{c_{3m}-1}, \quad (8.148)$$

where c_{im} $(i=1,2,3)$ is given by Equation (3.40). With regard to Equations (8.4), (8.6), (8.7), (8.11)–(8.18), (8.144), (8.145), we get

$$u'_{1m} = -\frac{1}{\zeta_m}\left\{p_p\left[\left(\frac{1}{2}+\ln r_c\right)r^{c_{3m}} - r_c^{c_{3m}}\left(\frac{1}{2}+\ln r\right)\right]\right.$$

$$+ \frac{u_{1c0}}{d}\left[\zeta_{1m} r^{c_{3m}} - \zeta_{2m}\left(\frac{1}{2} + \ln r\right)\right]\Bigg\}, \quad (8.149)$$

$$\varepsilon'_{11m} = -\frac{1}{\zeta_m}\Bigg\{p_p\left[c_{3m}\left(\frac{1}{2} + \ln r_c\right) r^{c_{3m}-1} - \frac{r_c^{c_{3m}}}{r}\right] + \frac{u_{1c0}}{d}\left(c_{3m}\zeta_{1m} r^{c_{3m}-1} - \frac{\zeta_{2m}}{r}\right)\Bigg\}, \quad (8.150)$$

$$\varepsilon'_{22m} = \varepsilon'_{33m} = -\frac{1}{\zeta_m}\Bigg\{p_p\left[\left(\frac{1}{2} + \ln r_c\right) r^{c_{3m}-1} - \frac{r_c^{c_{3m}}}{r}\left(\frac{1}{2} + \ln r\right)\right] + \frac{u_{1c0}}{d}\left[\zeta_{1m} r^{c_{3m}-1} - \frac{\zeta_{2m}}{r}\left(\frac{1}{2} + \ln r\right)\right]\Bigg\}, \quad (8.151)$$

$$\begin{aligned}\varepsilon'_{12m} = -\frac{1}{\zeta_m}\Bigg\{&\left[\left(\frac{1}{2} + \ln r_c\right) r^{c_{3m}-1} - \frac{r_c^{c_{3m}}}{r}\left(\frac{1}{2} + \ln r\right)\right]\left(\frac{\partial p_p}{\partial} - \frac{p_p}{\zeta_m}\frac{\partial \zeta_m}{\partial}\right)\\ &+ p_p\left[\frac{r^{c_{3m}-1}}{r_c} - \frac{c_{3m} r_c^{c_{3m}-1}}{r}\left(\frac{1}{2} + \ln r\right)\right]\frac{\partial r_c}{\partial \varphi}\\ &- \frac{u_{1c0}}{d\zeta_m}\left[\zeta_{1m} r^{c_{3m}-1} - \frac{\zeta_{2m}}{r}\left(\frac{1}{2} + \ln r\right)\right]\frac{\partial \zeta_m}{\partial \varphi}\\ &+ \frac{u_{1c0}}{d}\left[r^{c_{3m}-1}\frac{\partial \zeta_{1m}}{\partial \varphi} - \frac{1}{r}\left(\frac{1}{2} + \ln r\right)\frac{\partial \zeta_{2m}}{\partial \varphi}\right]\Bigg\}, \end{aligned} \quad (8.152)$$

$$\begin{aligned}\varepsilon'_{13m} = -\frac{1}{\zeta_m}\Bigg\{&\left[\left(\frac{1}{2} + \ln r_c\right) r^{c_{3m}-1} - \frac{r_c^{c_{3m}}}{r}\left(\frac{1}{2} + \ln r\right)\right]\left(\frac{\partial p_p}{\partial} - \frac{p_p}{\zeta_m}\frac{\partial \zeta_m}{\partial}\right)\\ &+ p_p\left[\frac{r^{c_{3m}-1}}{r_c} - \frac{c_{3m} r_c^{c_{3m}-1}}{r}\left(\frac{1}{2} + \ln r\right)\right]\frac{\partial r_c}{\partial \nu}\\ &- \frac{u_{1c0}}{d\zeta_m}\left[\zeta_{1m} r^{c_{3m}-1} - \frac{\zeta_{2m}}{r}\left(\frac{1}{2} + \ln r\right)\right]\frac{\partial \zeta_m}{\partial \nu}\\ &+ \frac{u_{1c0}}{d}\left[r^{c_{3m}-1}\frac{\partial \zeta_{1m}}{\partial \nu} - \frac{1}{r}\left(\frac{1}{2} + \ln r\right)\frac{\partial \zeta_{2m}}{\partial \nu}\right]\Bigg\}, \end{aligned} \quad (8.153)$$

$$\sigma'_{11m} = -\frac{1}{\zeta_m}\left(p_p r^{c_{3m}-1}\left\{[c_{3m}(c_{1m}+c_{2m})-2c_{2m}]\left(\frac{1}{2}+\ln r_c\right)\right.\right.$$
$$\left.-\frac{r_c}{r}(c_{1m}-2c_{2m}\ln r)\right\}$$
$$+\frac{u_{1c0}}{d}\left\{\zeta_{1m} r^{c_{3m}-1}[c_{3m}(c_{1m}+c_{2m})-2c_{2m}]\right.$$
$$\left.\left.-\frac{\zeta_{2m}}{r}(c_{1m}-2c_{2m}\ln r)\right\}\right), \tag{8.154}$$

$$\sigma'_{22m} = \sigma'_{33m} = -\frac{1}{\zeta_m}\left\{p_p r^{c_{3m}-1}\left[(c_{1m}-c_{2m}c_{3m})\left(\frac{1}{2}+\ln r_c\right)\right.\right.$$
$$\left.-\frac{r_c}{r}\left(\frac{c_{1m}-2c_{2m}}{2}+c_{1m}\ln r\right)\right]$$
$$+\frac{u_{1c0}}{d}\left[\zeta_{1m} r^{c_{3m}-1}(c_{1m}-c_{2m}c_{3m})\right.$$
$$\left.\left.-\frac{\zeta_{2m}}{r}\left(\frac{c_{1m}-2c_{2m}}{2}+c_{1m}\ln r\right)\right]\right\}, \tag{8.155}$$

$$\sigma'_{12m} = -\frac{1}{s_{44m}\zeta_m}\left\{\left[\left(\frac{1}{2}+\ln r_c\right)r^{c_{3m}-1}-\frac{r_c^{c_{3m}}}{r}\left(\frac{1}{2}+\ln r\right)\right]\right.$$
$$\times\left(\frac{\partial p_p}{\partial}-\frac{p_p}{\zeta_m}\frac{\partial \zeta_m}{\partial}\right)$$
$$+p_p\left[\frac{r^{c_{3m}-1}}{r_c}-\frac{c_{3m}r_c^{c_{3m}-1}}{r}\left(\frac{1}{2}+\ln r\right)\right]\frac{\partial r_c}{\partial \varphi}$$
$$-\frac{u_{1c0}}{d\zeta_m}\left[\zeta_{1m}r^{c_{3m}-1}-\frac{\zeta_{2m}}{r}\left(\frac{1}{2}+\ln r\right)\right]\frac{\partial \zeta_m}{\partial \varphi}$$
$$\left.+\frac{u_{1c0}}{d}\left[r^{c_{3m}-1}\frac{\partial \zeta_{1m}}{\partial \varphi}-\frac{1}{r}\left(\frac{1}{2}+\ln r\right)\frac{\partial \zeta_{2m}}{\partial \varphi}\right]\right\}, \tag{8.156}$$

$$\sigma'_{13m} = -\frac{1}{s_{44m}\zeta_m}\left\{\left[\left(\frac{1}{2}+\ln r_c\right)r^{c_{3m}-1}-\frac{r_c^{c_{3m}}}{r}\left(\frac{1}{2}+\ln r\right)\right]\right.$$

$$\times \left(\frac{\partial p_p}{\partial} - \frac{p_p}{\zeta_m} \frac{\partial \zeta_m}{\partial} \right)$$

$$+ p_p \left[\frac{r^{c_{3m}-1}}{r_c} - \frac{c_{3m} r_c^{c_{3m}-1}}{r} \left(\frac{1}{2} + \ln r \right) \right] \frac{\partial r_c}{\partial \nu}$$

$$- \frac{u_{1c0}}{d \zeta_m} \left[\zeta_{1m} r^{c_{3m}-1} - \frac{\zeta_{2m}}{r} \left(\frac{1}{2} + \ln r \right) \right] \frac{\partial \zeta_m}{\partial \nu}$$

$$+ \frac{u_{1c0}}{d} \left[r^{c_{3m}-1} \frac{\partial \zeta_{1m}}{\partial \nu} - \frac{1}{r} \left(\frac{1}{2} + \ln r \right) \frac{\partial \zeta_{2m}}{\partial \nu} \right] \Bigg\}, \quad (8.157)$$

where s_{44m} is given by Equation (3.35).

8.4.2. Cell Matrix around Void

As presented in Section 5.3, the absolute values $|u'_{1v}|$, $\left|\varepsilon'_{ijv}\right|$, $\left|\sigma'_{ijv}\right|$ $(i, j = 1,2,3)$ are required to represent increasing functions of $r \in \langle R_v, r_c \rangle$ (see Equations (3.19)–(3.23), (3.36)–(3.39)). These increasing functions exhibit maximum values on the cell surface, i.e., for $r = r_c$ (see Equations (3.15)–(3.18)). These increasing courses are ensured by the integration constants $C_{2v} \neq 0$, $C_{3v} \neq 0$, which are determined by the mandatory boundary conditions (5.5), (5.6), where $C_{1v} = 0$. With regard to Equations (5.5), (5.6), (8.4), (8.6), (8.7), (8.15), we get

$$C_{2v} = \frac{u'_{1c} \chi_{1v}}{\xi_{2v} \chi_v r_c} = \frac{u_{1c0} \chi_{1v}}{d \xi_{2v} \chi_v}, \quad (8.158)$$

$$C_{3v} = -\frac{u'_{1c} \chi_{2v}}{\xi_{3v} \chi_v r_c} = -\frac{u_{1c0} \chi_{2v}}{d \xi_{3v} \chi_v}, \quad (8.159)$$

where $u'_{1c}/r_c = u_{1c0}/d$ (see Equations (3.15)–(3.18), (5.4)); and r_c, ξ_{iv} $(i = 2,3)$ are given by Equations (3.15)–(3.18), (8.9), (8.10), respectively. The displacement u_{1c0} along the axis x_j $(j = 1,2,3)$ on the cell surface I, II (see Equation (5.4)), i.e., for $r = d/2$ (see Figure 2.1), is determined in Section 8.4.4. The coefficients χ_{iv} $(i = 1,2)$, χ_v have the forms

$$\chi_{1v} = \frac{r_c}{R_v} \left(c_{1m} - 2 c_{2m} \ln R_v \right), \quad (8.160)$$

$$\chi_{2m} = r_c \left[c_{3m} \left(c_{1m} + c_{2m} \right) - 2 c_{2m} \right] R_v^{c_{3m}-1}, \quad (8.161)$$

$$\chi_v = \frac{\chi_{2v}}{r_c}\left(\frac{1}{2} + \ln r_c\right) - \chi_{1v} r_c^{c_{3m}-1}, \tag{8.162}$$

where c_{im} $(i = 1,2)$ is given by Equation (3.40). With regard to Equations (8.4), (8.6), (8.7), (8.11)–(8.18), (8.158), (8.159), we get

$$u'_{1m} = \frac{u_{1c0}}{d\chi_v}\left[\chi_{1v} r^{c_{3m}} - \chi_{2v}\left(\frac{1}{2} + \ln r\right)\right], \tag{8.163}$$

$$\varepsilon'_{11m} = \frac{u_{1c0}}{d\chi_v}\left(c_{3m}\chi_{1v} r^{c_{3m}-1} - \frac{\chi_{2v}}{r}\right), \tag{8.164}$$

$$\varepsilon'_{22m} = \varepsilon'_{33m} = \frac{u_{1c0}}{d\chi_v}\left[\chi_{1v} r^{c_{3m}-1} - \frac{\chi_{2v}}{r}\left(\frac{1}{2} + \ln r\right)\right], \tag{8.165}$$

$$\varepsilon'_{12m} = \frac{u_{1c0}}{d\chi_v}\left\{\left[r^{c_{3m}-1}\frac{\partial \chi_{1v}}{\partial \varphi} - \frac{1}{r}\left(\frac{1}{2} + \ln r\right)\frac{\partial \chi_{2v}}{\partial \varphi}\right]\right.$$
$$\left. - \frac{1}{\chi_v}\left[\chi_{1v} r^{c_{3m}-1} - \frac{\chi_{2v}}{r}\left(\frac{1}{2} + \ln r\right)\right]\frac{\partial \chi_v}{\partial \varphi}\right\}, \tag{8.166}$$

$$\varepsilon'_{13m} = \frac{u_{1c0}}{d\chi_v}\left\{\left[r^{c_{3m}-1}\frac{\partial \chi_{1v}}{\partial \nu} - \frac{1}{r}\left(\frac{1}{2} + \ln r\right)\frac{\partial \chi_{2v}}{\partial \nu}\right]\right.$$
$$\left. - \frac{1}{\chi_v}\left[\chi_{1v} r^{c_{3m}-1} - \frac{\chi_{2v}}{r}\left(\frac{1}{2} + \ln r\right)\right]\frac{\partial \chi_v}{\partial \nu}\right\}, \tag{8.167}$$

$$\sigma'_{11m} = \frac{u_{1c0}}{d\chi_v}\left\{\chi_{1v} r^{c_{3m}-1}\left[c_{3m}(c_{1m} + c_{2m}) - 2c_{2m}\right] - \frac{\chi_{2v}}{r}(c_{1m} - 2c_{2m}\ln r)\right\}, \tag{8.168}$$

$$\sigma'_{22m} = \sigma'_{33m} = \frac{u_{1c0}}{d\chi_v}\left[\chi_{1v} r^{c_{3m}-1}(c_{1m} - c_{2m}c_{3m})\right.$$
$$\left. - \frac{\chi_{2v}}{r}\left(\frac{c_{1m} - 2c_{2m}}{2} + c_{1m}\ln r\right)\right], \tag{8.169}$$

$$\sigma'_{12m} = \frac{u_{1c0}}{d\,s_{44m}\chi_v}\left\{\left[r^{c_{3m}-1}\frac{\partial \chi_{1v}}{\partial \varphi} - \frac{1}{r}\left(\frac{1}{2} + \ln r\right)\frac{\partial \chi_{2v}}{\partial \varphi}\right]\right.$$

$$- \frac{1}{\chi_v} \left[\chi_{1v} r^{c_{3m}-1} - \frac{\chi_{2v}}{r} \left(\frac{1}{2} + \ln r \right) \right] \frac{\partial \chi_v}{\partial \varphi} \Bigg\}, \quad (8.170)$$

$$\sigma'_{13m} = \frac{u_{1c0}}{d\, s_{44m} \chi_v} \Bigg\{ \left[r^{c_{3m}-1} \frac{\partial \chi_{1v}}{\partial \nu} - \frac{1}{r} \left(\frac{1}{2} + \ln r \right) \frac{\partial \chi_{2v}}{\partial \nu} \right] - \frac{1}{\chi_v} \left[\chi_{1v} r^{c_{3m}-1} - \frac{\chi_{2v}}{r} \left(\frac{1}{2} + \ln r \right) \right] \frac{\partial \chi_v}{\partial \nu} \Bigg\}, \quad (8.171)$$

where s_{44m} is given by Equation (3.35).

8.4.3. Radial Stress p_p

With regard to Equations (5.12), (6.25), (8.151), the radial stress p_p, which acts at the particle-matrix boundary, i.e., for $r = R_p$ (see Figure 2.1a), is given by Equations (8.60)–(8.63). The coefficients ζ_{3m}, ζ_{4m}, which are included in Equations (8.60), (8.61), have the forms

$$\zeta_{3m} = \frac{1}{\zeta_m} \left[\left(\frac{1}{2} + \ln r_c \right) R_p^{c_{3m}-1} - \frac{r_c^{c_{3m}}}{R_p} \left(\frac{1}{2} + \ln R_p \right) \right], \quad (8.172)$$

$$\zeta_{4m} = \frac{1}{\zeta_m} \left[\zeta_{1m} R_p^{c_{3m}-1} - \frac{\zeta_{2m}}{R_p} \left(\frac{1}{2} + \ln R_p \right) \right], \quad (8.173)$$

where r_c, c_{3m}, ζ_{im} $(i = 1,2)$, ζ_m are given by Equations (3.15)–(3.18), (3.40), (8.146)–(8.148), respectively.

8.4.4. Displacement u_{1c0}

With regard to Equations (8.60), (8.144), (8.145), (8.158), (8.159), the integration constant C_{iq} $(i = 2,3;\ q = m,v)$ has the form

$$C_{im} = \frac{1}{\xi_{im}} \left[\frac{u_{1c0} \zeta_{i1m}}{d} + \zeta_{i2m} (\beta_m - \beta_p) \right], \quad i = 2,3, \quad (8.174)$$

$$C_{iv} = \frac{u_{1c0} \zeta_{iv}}{d\, \xi_{iv}}, \quad i = 2,3, \quad (8.175)$$

where ξ_{im}, ξ_{iv} $(i = 2,3)$ are given by Equations (8.9), (8.10), and β_q $(q = p,m)$ is determined in Section 4.2. The coefficients ζ_{i1m}, ζ_{i2m}, ζ_{iv} $(i = 2,3)$ are derived as

$$\zeta_{21m} = -\frac{1}{\zeta_m}\left[\zeta_{1m} + \zeta_{pm}\zeta_{4m}\left(\frac{1}{2} + \ln r_c\right)\right], \tag{8.176}$$

$$\zeta_{31m} = \frac{1}{\zeta_m}\left(\zeta_{2m} + \zeta_{pm}\zeta_{4m}r^{c_{3m}}\right), \tag{8.177}$$

$$\zeta_{22m} = -\frac{\zeta_{pm}}{\zeta_m}\left(\frac{1}{2} + \ln r_c\right), \tag{8.178}$$

$$\zeta_{32m} = \frac{\zeta_{pm}r^{c_{3m}}}{\zeta_m}, \tag{8.179}$$

$$\zeta_{2v} = \frac{\chi_{1v}}{\chi_v}, \tag{8.180}$$

$$\zeta_{3v} = -\frac{\chi_{2v}}{\chi_v}, \tag{8.181}$$

where r_c, c_{3m}, ζ_{im}, ζ_m, χ_{iv} $(i=1,2)$, χ_v and ζ_{pm}, ζ_{4m} are given by Equations (3.15)–(3.18), (3.40), (8.146)–(8.148), (8.160)–(8.162) and (8.61), (8.173), respectively. The coefficients ζ_{3m}, ζ_{4m}, which are included in ζ_{pm} (see Equation (8.61)), are given by Equations (8.172), (8.173), respectively.

With regard to Equations (8.19), (8.23)–(8.25), (8.29)–(8.31), (8.174), (8.175), the elastic energy density w_q $(q=m,v)$ is given by Equations (8.72), (8.73). The coefficients $\omega_{im}^{(12)}$ $(i=1,2,3)$, $\omega_v^{(12)}$, which are included in Equations (8.72), (8.73), are replaced by $\omega_{im}^{(23)}$ $(i=1,2,3)$, $\omega_v^{(23)}$, respectively. The coefficients $\omega_{im}^{(23)}$ $(i=1,2,3)$, $\omega_v^{(23)}$ are derived as

$$\begin{aligned}\omega_{1m}^{(23)} &= \zeta_{21m}^2 u_{22m}^{(r)} + \zeta_{31m}^2 u_{33m}^{(r)} + \zeta_{21m}\zeta_{31m}u_{23m}^{(r)} \\ &+ \frac{u_{22m}^{(\varphi\nu)}}{s_{44m}}\left[\left(\frac{\partial\zeta_{21m}}{\partial\varphi}\right)^2 + \left(\frac{\partial\zeta_{21m}}{\partial\nu}\right)^2\right] \\ &+ \frac{u_{33m}^{(\varphi\nu)}}{s_{44m}}\left[\left(\frac{\partial\zeta_{31m}}{\partial\varphi}\right)^2 + \left(\frac{\partial\zeta_{31m}}{\partial\nu}\right)^2\right] \\ &+ \frac{u_{23m}^{(\varphi\nu)}}{s_{44m}}\left[\frac{\partial\zeta_{21m}}{\partial\varphi}\frac{\partial\zeta_{31m}}{\partial\varphi} + \frac{\partial\zeta_{21m}}{\partial\nu}\frac{\partial\zeta_{31m}}{\partial\nu}\right],\end{aligned} \tag{8.182}$$

$$\omega_{2m}^{(23)} = 2\left[\zeta_{21m}\,\zeta_{22m}\,u_{22m}^{(r)} + \zeta_{31m}\,\zeta_{32m}\,u_{33m}^{(r)}\right] + (\zeta_{21m}\,\zeta_{32m} + \zeta_{22m}\,\zeta_{31m})\,u_{23m}^{(r)}$$
$$+ \frac{2\,u_{22m}^{(\varphi\nu)}}{s_{44m}}\left(\frac{\partial\zeta_{21m}}{\partial\varphi}\frac{\partial\zeta_{22m}}{\partial\varphi} + \frac{\partial\zeta_{21m}}{\partial\nu}\frac{\partial\zeta_{22m}}{\partial\nu}\right)$$
$$+ \frac{2\,u_{33m}^{(\varphi\nu)}}{s_{44m}}\left(\frac{\partial\zeta_{31m}}{\partial\varphi}\frac{\partial\zeta_{32m}}{\partial\varphi} + \frac{\partial\zeta_{31m}}{\partial\nu}\frac{\partial\zeta_{32m}}{\partial\nu}\right)$$
$$+ \frac{u_{23m}^{(\varphi\nu)}}{s_{44m}}\left(\frac{\partial\zeta_{21m}}{\partial\varphi}\frac{\partial\zeta_{32m}}{\partial\varphi} + \frac{\partial\zeta_{22m}}{\partial\varphi}\frac{\partial\zeta_{31m}}{\partial\varphi}\right.$$
$$\left. + \frac{\partial\zeta_{21m}}{\partial\nu}\frac{\partial\zeta_{32m}}{\partial\nu} + \frac{\partial\zeta_{22m}}{\partial\nu}\frac{\partial\zeta_{31m}}{\partial\nu}\right), \tag{8.183}$$

$$\omega_{3m}^{(23)} = \zeta_{22m}^2\,u_{22m}^{(r)} + \zeta_{32m}^2\,u_{33m}^{(r)} + \zeta_{22m}\,\zeta_{32m}\,u_{23m}^{(r)}$$
$$+ \frac{u_{22m}^{(\varphi\nu)}}{s_{44m}}\left[\left(\frac{\partial\zeta_{22m}}{\partial\varphi}\right)^2 + \left(\frac{\partial\zeta_{22m}}{\partial\nu}\right)^2\right]$$
$$+ \frac{u_{33m}^{(\varphi\nu)}}{s_{44m}}\left[\left(\frac{\partial\zeta_{32m}}{\partial\varphi}\right)^2 + \left(\frac{\partial\zeta_{32m}}{\partial\nu}\right)^2\right]$$
$$+ \frac{u_{23m}^{(\varphi\nu)}}{s_{44m}}\left[\frac{\partial\zeta_{22m}}{\partial\varphi}\frac{\partial\zeta_{32m}}{\partial\varphi} + \frac{\partial\zeta_{22m}}{\partial\nu}\frac{\partial\zeta_{32m}}{\partial\nu}\right], \tag{8.184}$$

$$\omega_{v}^{(23)} = \zeta_{2v}^2\,u_{22m}^{(r)} + \zeta_{3v}^2\,u_{33m}^{(r)} + \zeta_{2v}\,\zeta_{3v}\,u_{23m}^{(r)}$$
$$+ \frac{u_{22m}^{(\varphi\nu)}}{s_{44m}}\left[\left(\frac{\partial\zeta_{2v}}{\partial\varphi}\right)^2 + \left(\frac{\partial\zeta_{2v}}{\partial\nu}\right)^2\right]$$
$$+ \frac{u_{33m}^{(\varphi\nu)}}{s_{44m}}\left[\left(\frac{\partial\zeta_{3v}}{\partial\varphi}\right)^2 + \left(\frac{\partial\zeta_{3v}}{\partial\nu}\right)^2\right]$$
$$+ \frac{u_{23m}^{(\varphi\nu)}}{s_{44m}}\left[\frac{\partial\zeta_{2v}}{\partial\varphi}\frac{\partial\zeta_{3v}}{\partial\varphi} + \frac{\partial\zeta_{2v}}{\partial\nu}\frac{\partial\zeta_{3v}}{\partial\nu}\right], \tag{8.185}$$

where s_{44m} is given by Equation (3.35).

With regard to Equations (3.47), (8.72), the elastic energy W_m, which is accumulated in the cell matrix *I*, is given by Equation (8.78). With regard to Equations (8.72), (8.182)–(8.185), the coefficient $\Omega_{im}^{(12)}$ $(i=1,2,3)$, which is included in Equation (7.76), is replaced by $\Omega_{im}^{(23)}$ $(i=1,2,3)$. The coefficient $\Omega_{im}^{(23)}$ $(i=1,2,3)$ has the form

$$\begin{aligned}
\Omega_{1m}^{(23)} &= \int_0^{\pi/2}\int_0^{\pi/2}\int_{R_p}^{r_c} \omega_{1m}^{(23)} r^2\, dr\, d\varphi\, d\nu \\
&= \int_0^{\pi/2}\int_0^{\pi/2} \left\{ \zeta_{21m}^2 U_{22m}^{(r)} + \zeta_{31m}^2 U_{33m}^{(r)} + \zeta_{21m}\zeta_{31m} U_{23m}^{(r)} \right. \\
&\qquad + \frac{U_{22m}^{(\varphi\nu)}}{s_{44m}} \left[\left(\frac{\partial \zeta_{21m}}{\partial \varphi} \right)^2 + \left(\frac{\partial \zeta_{21m}}{\partial \nu} \right)^2 \right] \\
&\qquad + \frac{U_{33m}^{(\varphi\nu)}}{s_{44m}} \left[\left(\frac{\partial \zeta_{31m}}{\partial \varphi} \right)^2 + \left(\frac{\partial \zeta_{31m}}{\partial \nu} \right)^2 \right] \\
&\qquad \left. + \frac{U_{23m}^{(\varphi\nu)}}{s_{44m}} \left[\frac{\partial \zeta_{21m}}{\partial \varphi} \frac{\partial \zeta_{31m}}{\partial \varphi} + \frac{\partial \zeta_{21m}}{\partial \nu} \frac{\partial \zeta_{31m}}{\partial \nu} \right] \right\} d\varphi\, d\nu, \quad (8.186)
\end{aligned}$$

$$\begin{aligned}
\Omega_{2m}^{(23)} &= \int_0^{\pi/2}\int_0^{\pi/2}\int_{R_p}^{r_c} \omega_{2m}^{(23)} r^2\, dr\, d\varphi\, d\nu \\
&= \int_0^{\pi/2}\int_0^{\pi/2} \left\{ 2 \left[\zeta_{21m}\zeta_{22m} U_{22m}^{(r)} + \zeta_{31m}\zeta_{32m} U_{33m}^{(r)} \right] \right. \\
&\qquad + \left(\zeta_{21m}\zeta_{32m} + \zeta_{22m}\zeta_{31m} \right) U_{23m}^{(r)} \\
&\qquad + \frac{2U_{22m}^{(\varphi\nu)}}{s_{44m}} \left(\frac{\partial \zeta_{21m}}{\partial \varphi} \frac{\partial \zeta_{22m}}{\partial \varphi} + \frac{\partial \zeta_{21m}}{\partial \nu} \frac{\partial \zeta_{22m}}{\partial \nu} \right) \\
&\qquad + \frac{2U_{33m}^{(\varphi\nu)}}{s_{44m}} \left(\frac{\partial \zeta_{31m}}{\partial \varphi} \frac{\partial \zeta_{32m}}{\partial \varphi} + \frac{\partial \zeta_{31m}}{\partial \nu} \frac{\partial \zeta_{32m}}{\partial \nu} \right) \\
&\qquad + \frac{U_{23m}^{(\varphi\nu)}}{s_{44m}} \left(\frac{\partial \zeta_{21m}}{\partial \varphi} \frac{\partial \zeta_{32m}}{\partial \varphi} + \frac{\partial \zeta_{22m}}{\partial \varphi} \frac{\partial \zeta_{31m}}{\partial \varphi} \right)
\end{aligned}$$

$$+\frac{U_{23m}^{(\varphi\nu)}}{s_{44m}}\left(\frac{\partial\zeta_{21m}}{\partial\nu}\frac{\partial\zeta_{32m}}{\partial\nu}+\frac{\partial\zeta_{22m}}{\partial\nu}\frac{\partial\zeta_{31m}}{\partial\nu}\right)\Bigg\}\,d\varphi\,d\nu, \quad (8.187)$$

$$\Omega_{3m}^{(23)} = \int_0^{\pi/2}\int_0^{\pi/2}\int_{R_p}^{r_c}\omega_{3m}^{(23)}\,r^2\,dr\,d\varphi\,d\nu$$
$$= \int_0^{\pi/2}\int_0^{\pi/2}\Bigg\{\zeta_{22m}^2 U_{22m}^{(r)} + \zeta_{32m}^2 U_{33m}^{(r)} + \zeta_{22m}\zeta_{32m}U_{23m}^{(r)}$$
$$+\frac{U_{22m}^{(\varphi\nu)}}{s_{44m}}\left[\left(\frac{\partial\zeta_{22m}}{\partial\varphi}\right)^2+\left(\frac{\partial\zeta_{22m}}{\partial\nu}\right)^2\right]$$
$$+\frac{U_{33m}^{(\varphi\nu)}}{s_{44m}}\left[\left(\frac{\partial\zeta_{32m}}{\partial\varphi}\right)^2+\left(\frac{\partial\zeta_{32m}}{\partial\nu}\right)^2\right]$$
$$+\frac{U_{23m}^{(\varphi\nu)}}{s_{44m}}\left[\frac{\partial\zeta_{22m}}{\partial\varphi}\frac{\partial\zeta_{32m}}{\partial\varphi}+\frac{\partial\zeta_{22m}}{\partial\nu}\frac{\partial\zeta_{32m}}{\partial\nu}\right]\Bigg\}\,d\varphi\,d\nu, \quad (8.188)$$

where $U_{ijm}^{(x)}$ $(i,j=2,3;\ x=r,\varphi\nu)$, $U_{iim}^{(r)}$, $U_{iim}^{(\varphi\nu)}$ are given by Equations (8.91), (8.85), (8.88), (8.140), (8.142). With regard to Equations (8.24), (8.30), we get

$$U_{23m}^{(r)} = \frac{2c_{1m}(1-c_{3m})}{c_{3m}+1}$$
$$\times\left[r_c^{c_{3m}+1}\left(\ln r_c - \frac{1}{c_{3m}+1}\right) - R_p^{c_{3m}+1}\left(\ln R_p - \frac{1}{c_{3m}+1}\right)\right]$$
$$+\frac{c_{2m}c_{3m}-c_{1m}}{c_{3m}+1}\left(r_c^{c_{3m}+1}-R_p^{c_{3m}+1}\right), \quad (8.189)$$

$$U_{23m}^{(\varphi\nu)} = \frac{2}{c_{3m}+1}\left[r_c^{c_{3m}+1}\left(\ln r_c - \frac{1}{c_{3m}+1}\right) - R_p^{c_{3m}+1}\left(\ln R_p - \frac{1}{c_{3m}+1}\right)\right]$$
$$+\frac{1}{c_{3m}+1}\left(r_c^{c_{3m}+1}-R_p^{c_{3m}+1}\right), \quad (8.190)$$

where r_c, c_{im} $(i=1,3)$ are given by Equations (3.15)–(3.18), (3.40), respectively.

With regard to Equations (3.48), (8.73), the elastic energy W_v, which is accumulated in the cell matrix *II*, is given by Equation (8.89). With regard to Equations (8.73), (8.185), the coefficient $\Omega_v^{(12)}$, which is included in Equation (8.89), is replaced by $\Omega_v^{(23)}$. The coefficient $\Omega_v^{(23)}$ has the form

$$\begin{aligned}
\Omega_v^{(23)} &= \int_0^{\pi/2}\int_0^{\pi/2}\int_{R_p}^{r_c} \omega_v^{(23)} r^2\, dr\, d\varphi\, d\nu \\
&= \int_0^{\pi/2}\int_0^{\pi/2} \left\{ \zeta_{2v}^2 U_{22v}^{(r)} + \zeta_{3v}^2 U_{33v}^{(r)} \right. \\
&\quad + \frac{U_{22v}^{(\varphi\nu)}}{s_{44m}} \left[\left(\frac{\partial \zeta_{2v}}{\partial \varphi}\right)^2 + \left(\frac{\partial \zeta_{2v}}{\partial \nu}\right)^2 \right] \\
&\quad + \frac{U_{33v}^{(\varphi\nu)}}{s_{44m}} \left[\left(\frac{\partial \zeta_{3v}}{\partial \varphi}\right)^2 + \left(\frac{\partial \zeta_{3v}}{\partial \nu}\right)^2 \right] \\
&\quad \left. + \frac{U_{23v}^{(\varphi\nu)}}{s_{44m}} \left[\frac{\partial \zeta_{2v}}{\partial \varphi}\frac{\partial \zeta_{3v}}{\partial \varphi} + \frac{\partial \zeta_{2v}}{\partial \nu}\frac{\partial \zeta_{3v}}{\partial \nu} \right] \right\} d\varphi\, d\nu. \qquad (8.191)
\end{aligned}$$

With regard to Equations (8.82), (8.91), $U_{ijv}^{(\varphi\nu)}$ $(i, j = 2,3;\ x = r,\ \varphi\nu)$ is given by Equations (8.85), (8.88), (8.140), (8.142), (8.190). In case of $U_{ijv}^{(x)}$, the radius R_p, which is included in Equations (8.85), (8.88), (8.140), (8.142), (8.190), is replaced by R_v.

With regard to Equations (5.13), (6.81), (8.78), (8.89), the displacement u_{1c0} along the axis x_i $(i = 1,2,3)$ on the cubic cell surface *I*, *II*, i.e., for $r = d/2$ (see Figure 2.1a), is given by Equations (6.89), (8.92), (8.93).

The coefficients $\Omega_{1m}^{(12)}$, $\Omega_{2m}^{(12)}$, $\Omega_v^{(12)}$, which are included in Equations (8.92), (8.93), are replaced by $\Omega_{1m}^{(23)}$, $\Omega_{2m}^{(23)}$, $\Omega_v^{(23)}$ (see Equations (8.186), (8.187), (8.191)), respectively.

8.5. Integration Constants $C_{1q} \neq 0$, $C_{2q} \neq 0$, $C_{3q} \neq 0$

8.5.1. Cell Matrix Around Particle

As presented above (see page 105), $u_{1rm} = u_{1rm}(r)$ is an increasing function of the variable r for $r < e^{1/3} \approx 1.39561$ m (see Equation (8.5)), $u_{2rm} = u_{2rm}(r)$ is a decreasing function of the variable r due to $c_{3m} < 0$ (see Equations (3.40), (8.6)), and $u_{3rm} = u_{3rm}(r)$ is an increasing function of the variable r (see Equation (8.7)).

As presented in Section 5.2, the absolute values $|u'_{1m}|$, $\left|\varepsilon'_{ijm}\right|$, $\left|\sigma'_{ijm}\right|$ $(i, j = 1,2,3)$ are required to represent decreasing functions of $r \in \langle R_p, r_c\rangle$ (see Equations (3.19)–(3.23), (3.36)–(3.39)). These decreasing functions exhibit maximum values on the particle-matrix boundary, i.e., for $r = R_p$. These decreasing courses are ensured by the integration constants $C_{1m} \neq 0$, $C_{2m} \neq 0$, $C_{3m} \neq 0$, which are determined by the mandatory and additional boundary conditions (5.3), (5.4) and (5.7), respectively. With regard to Equations (5.3), (5.4), (5.7), (8.4)–(8.7), (8.11), (8.15), we get

$$C_{im} = \frac{1}{\xi_{im}\zeta_m}\left[p_p\,\zeta_m^{(i1)} + \frac{u'_{1c}\,\zeta_m^{(i2)}}{r_c}\right] = \frac{1}{\xi_{im}\zeta_m}\left[p_p\,\zeta_m^{(i1)} + \frac{u_{1c0}\,\zeta_m^{(i2)}}{d}\right],$$
$$i = 1,2,3, \tag{8.192}$$

where $u'_{1c}/r_c = u_{1c0}/d$ (see Equations (3.15)–(3.18), (5.4)); and r_c, ξ_{im} $(i = 1,2,3)$ are given by Equations (3.15)–(3.18), (7.8)–(7.10), respectively. The radial stress p_p, which acts at the particle-matrix boundary, i.e., for $r = R_p$ (see Figure 2.1a), is determined in Section 7.5.3. The displacement u_{1c0} along the axis x_i $(i = 1,2,3)$ on the cell surface I, II (see Equation (5.4)), i.e., for $r = d/2$ (see Figure 2.1), is determined in Section 7.5.4. The coefficients $\zeta_m^{(i1)}$, $\zeta_m^{(i2)}$ $(i = 1,2,3)$, ζ_m have the forms

$$\zeta_m^{(11)} = c_{3m}\left(\frac{1}{2} + \ln r_c\right)\left(r_c + u'_{1c}\right)^{c_{3m}-1} - \frac{r_c^{c_{3m}}}{r_c + u'_{1c}}, \tag{8.193}$$

$$\zeta_m^{(21)} = \frac{r_c}{r_c + u'_{1c}}\left(\frac{4}{3} - \ln r_c\right) - \left(\frac{1}{2} + \ln r_c\right)\left[\frac{1}{3} - \ln\left(r_c + u'_{1c}\right)\right], \tag{8.194}$$

$$\zeta_m^{(31)} = r_c\left\{r_c^{c_{3m}-1}\left[\frac{1}{3} - \ln\left(r_c + u'_{1c}\right)\right] - c_{3m}\left(\frac{4}{3} - \ln r_c\right)\left(r_c + u'_{1c}\right)^{c_{3m}-1}\right\}, \tag{8.195}$$

$$\zeta_m^{(12)} = r_c \left\{ c_{3m} \left(c_{1m} - 2\, c_{2m} \ln R_p \right) \frac{\left(r_c + u'_{1c} \right)^{c_{3m}-1}}{R_p} \right.$$
$$\left. - \left[\left(c_{1m} + c_{2m} \right) c_{3m} - 2\, c_{2m} \right] \frac{R_p^{c_{3m}-1}}{r_c + u'_{1c}} \right\}, \tag{8.196}$$

$$\zeta_m^{(22)} = r_c \left\{ \frac{1}{r_c + u'_{1c}} \left[\frac{c_{1m} - 7\, c_{2m}}{3} - \left(c_{1m} - c_{2m} \right) \ln R_p \right] \right.$$
$$\left. - \frac{c_{1m} - 2\, c_{2m} \ln R_p}{R_p} \left[\frac{1}{3} - \ln \left(r_c + u'_{1c} \right) \right] \right\} \tag{8.197}$$

$$\zeta_m^{(32)} = r_c \left\{ \left[\left(c_{1m} + c_{2m} \right) c_{3m} - 2\, c_{2m} \right] \left[\frac{1}{3} - \ln \left(r_c + u'_{1c} \right) \right] R_p^{c_{3m}-1} \right.$$
$$\left. - c_{3m} \left[\frac{c_{1m} - 7\, c_{2m}}{3} - \left(c_{1m} - c_{2m} \right) \ln R_p \right] \left(r_c + u'_{1c} \right)^{c_{3m}-1} \right\}, \tag{8.198}$$

$$\zeta_m = \frac{r_c^{c_{3m}}}{r_c + u'_{1c}} \left[\frac{c_{1m} - 7\, c_{2m}}{3} - \left(c_{1m} - c_{2m} \right) \ln R_p \right]$$
$$+ \left[\left(c_{1m} + c_{2m} \right) c_{3m} - 2\, c_{2m} \right] \left(\frac{1}{2} + \ln r_c \right) \left[\frac{1}{3} - \ln \left(r_c + u'_{1c} \right) \right] R_p^{c_{3m}-1}$$
$$+ \left(c_{1m} - 2\, c_{2m} \ln R_p \right) \left(\frac{4}{3} - \ln r_c \right) \frac{r_c\, c_{3m} \left(r_c + u'_{1c} \right)^{c_{3m}-1}}{R_p}$$
$$- \left\{ c_{3m} \left[\frac{c_{1m} - 7\, c_{2m}}{3} - \left(c_{1m} - c_{2m} \right) \ln R_p \right] \left(\frac{1}{2} + \ln r_c \right) \left(r_c + u'_{1c} \right)^{c_{3m}-1} \right.$$
$$+ \left[\left(c_{1m} + c_{2m} \right) c_{3m} - 2\, c_{2m} \right] \left(\frac{4}{3} - \ln r_c \right) \frac{r_c R_p^{c_{3m}-1}}{r_c + u'_{1c}}$$
$$\left. + \frac{\left(c_{1m} - 2\, c_{2m} \ln R_p \right) r_c^{c_{3m}}}{R_p} \left[\frac{1}{3} - \ln \left(r_c + u'_{1c} \right) \right] \right\}, \tag{8.199}$$

where c_{im} $(i = 1,2,3)$ is given by Equation (3.40). With regard to Equations (8.4)–(8.7), (8.11)–(8.18), (8.192), we get

$$u'_{1m} = \frac{1}{\zeta_m}\left\{p_p\left[\zeta_m^{(11)} r\left(\frac{4}{3} - \ln r\right) + \zeta_m^{(21)} r^{c_{3m}} + \zeta_m^{(31)}\left(\frac{1}{2} + \ln r\right)\right]\right.$$
$$\left. + \frac{u_{1c0}}{d}\left[\zeta_m^{(12)} r\left(\frac{4}{3} - \ln r\right) + \zeta_m^{(22)} r^{c_{3m}} + \zeta_m^{(32)}\left(\frac{1}{2} + \ln r\right)\right]\right\}, \tag{8.200}$$

$$\varepsilon'_{11m} = \frac{1}{\zeta_m}\left\{p_p\left[\zeta_m^{(11)}\left(\frac{1}{3} - \ln r\right) + \zeta_m^{(21)} c_{3m} r^{c_{3m}-1}\right]\right.$$
$$\left. + \frac{u_{1c0}}{d}\left[\zeta_m^{(12)}\left(\frac{1}{3} - \ln r\right) + \zeta_m^{(22)} c_{3m} r^{c_{3m}-1}\right]\right\}, \tag{8.201}$$

$$\varepsilon'_{22m} = \varepsilon'_{33m} = \frac{1}{\zeta_m}\left\{p_p\left[\zeta_m^{(11)}\left(\frac{4}{3} - \ln r\right) + \zeta_m^{(21)} r^{c_{3m}-1} + \frac{\zeta_m^{(31)}}{r}\left(\frac{1}{2} + \ln r\right)\right]\right.$$
$$\left. + \frac{u_{1c0}}{d}\left[\zeta_m^{(12)}\left(\frac{4}{3} - \ln r\right) + \zeta_m^{(22)} r^{c_{3m}-1} + \frac{\zeta_m^{(32)}}{r}\left(\frac{1}{2} + \ln r\right)\right]\right\}, \tag{8.202}$$

$$\varepsilon'_{12m} = \frac{1}{\zeta_m}\left(\left(\frac{4}{3} - \ln r\right)\left[\zeta_m^{(11)}\frac{\partial p_p}{\partial \varphi} + p_p\frac{\partial \zeta_m^{(11)}}{\partial \varphi} - \frac{p_p\,\zeta_m^{(11)}}{\zeta_m}\frac{\partial \zeta_m^{(11)}}{\partial \varphi}\right]\right.$$
$$+ r^{c_{3m}-1}\left[\zeta_m^{(21)}\frac{\partial p_p}{\partial \varphi} + p_p\frac{\partial \zeta_m^{(21)}}{\partial \varphi} - \frac{p_p\,\zeta_m^{(21)}}{\zeta_m}\frac{\partial \zeta_m^{(21)}}{\partial \varphi}\right]$$
$$+ \frac{1}{r}\left(\frac{1}{2} + \ln r\right)\left[\zeta_m^{(31)}\frac{\partial p_p}{\partial \varphi} + p_p\frac{\partial \zeta_m^{(31)}}{\partial \varphi} - \frac{p_p\,\zeta_m^{(31)}}{\zeta_m}\frac{\partial \zeta_m^{(31)}}{\partial \varphi}\right]$$
$$+ \frac{u_{1c0}}{d}\left\{\left(\frac{4}{3} - \ln r\right)\left[\frac{\partial \zeta_m^{(12)}}{\partial \varphi} - \frac{\zeta_m^{(12)}}{\zeta_m}\frac{\partial \zeta_m}{\partial \varphi}\right]\right.$$
$$+ r^{c_{3m}-1}\left[\frac{\partial \zeta_m^{(22)}}{\partial \varphi} - \frac{\zeta_m^{(22)}}{\zeta_m}\frac{\partial \zeta_m}{\partial \varphi}\right]$$

$$+ \frac{1}{r}\left(\frac{1}{2}+\ln r\right)\left[\frac{\partial\zeta_m^{(32)}}{\partial\varphi} - \frac{\zeta_m^{(32)}}{\zeta_m}\frac{\partial\zeta_m}{\partial\varphi}\right]\Bigg\}\Bigg), \tag{8.203}$$

$$\begin{aligned}\varepsilon'_{13m} = \frac{1}{\zeta_m}\Bigg(&\left(\frac{4}{3}-\ln r\right)\left[\zeta_m^{(11)}\frac{\partial p_p}{\partial\nu} + p_p\frac{\partial\zeta_m^{(11)}}{\partial\nu} - \frac{p_p\,\zeta_m^{(11)}}{\zeta_m}\frac{\partial\zeta_m^{(11)}}{\partial\nu}\right] \\ &+ r^{c_{3m}-1}\left[\zeta_m^{(21)}\frac{\partial p_p}{\partial\nu} + p_p\frac{\partial\zeta_m^{(21)}}{\partial\nu} - \frac{p_p\,\zeta_m^{(21)}}{\zeta_m}\frac{\partial\zeta_m^{(21)}}{\partial\nu}\right] \\ &+ \frac{1}{r}\left(\frac{1}{2}+\ln r\right)\left[\zeta_m^{(31)}\frac{\partial p_p}{\partial\nu} + p_p\frac{\partial\zeta_m^{(31)}}{\partial\nu} - \frac{p_p\,\zeta_m^{(31)}}{\zeta_m}\frac{\partial\zeta_m^{(31)}}{\partial\nu}\right] \\ &+ \frac{u_{1c0}}{d}\Bigg\{\left(\frac{4}{3}-\ln r\right)\left[\frac{\partial\zeta_m^{(12)}}{\partial\nu} - \frac{\zeta_m^{(12)}}{\zeta_m}\frac{\partial\zeta_m}{\partial\nu}\right] \\ &\quad + r^{c_{3m}-1}\left[\frac{\partial\zeta_m^{(22)}}{\partial\nu} - \frac{\zeta_m^{(22)}}{\zeta_m}\frac{\partial\zeta_m}{\partial\nu}\right] \\ &\quad + \frac{1}{r}\left(\frac{1}{2}+\ln r\right)\left[\frac{\partial\zeta_m^{(32)}}{\partial\nu} - \frac{\zeta_m^{(32)}}{\zeta_m}\frac{\partial\zeta_m}{\partial\nu}\right]\Bigg\}\Bigg),\end{aligned} \tag{8.204}$$

$$\begin{aligned}\sigma'_{11m} = \frac{1}{\zeta_m}\Bigg(&p_p\Bigg\{\zeta_m^{(11)}\left[\frac{c_{1m}-7\,c_{2m}}{3} - (c_{1m}-c_{2m})\ln r\right] \\ &+ \zeta_m^{(21)}\left[(c_{1m}+c_{2m})\,c_{3m} - 2\,c_{2m}\right]r^{c_{3m}-1} \\ &+ \frac{\zeta_m^{(31)}\,(c_{1m}-2\,c_{2m}\ln r)}{r}\Bigg\} \\ &+ \frac{u_{1c0}}{d}\Bigg\{\zeta_m^{(12)}\left[\frac{c_{1m}-7\,c_{2m}}{3} - (c_{1m}-c_{2m})\ln r\right] \\ &\quad + \zeta_m^{(22)}\left[(c_{1m}+c_{2m})\,c_{3m} - 2\,c_{2m}\right]r^{c_{3m}-1} \\ &\quad + \frac{\zeta_m^{(32)}\,(c_{1m}-2\,c_{2m}\ln r)}{r}\Bigg\}\Bigg),\end{aligned} \tag{8.205}$$

$$\sigma'_{22m} = \sigma'_{33m} = \frac{1}{\zeta_m}\Bigg(p_p\Bigg\{\zeta_m^{(11)}\left[\frac{4c_{1m}-c_{2m}}{3} - (c_{1m}-c_{2m})\ln r\right]$$

$$
\begin{aligned}
&+ \zeta_m^{(21)} \left(c_{1m} - c_{2m} c_{3m}\right) r^{c_{3m}-1} \\
&+ \zeta_m^{(31)} \left(\frac{c_{1m} - 2c_{2m}}{2} + c_{1m} \ln r\right)\Bigg\} \\
&+ \frac{u_{1c0}}{d} \left\{ \zeta_m^{(11)} \left[\frac{4c_{1m} - c_{2m}}{3} - \left(c_{1m} - c_{2m}\right) \ln r\right] \right. \\
&\quad + \zeta_m^{(21)} \left(c_{1m} - c_{2m} c_{3m}\right) r^{c_{3m}-1} \\
&\quad \left.\left. + \zeta_m^{(31)} \left(\frac{c_{1m} - 2c_{2m}}{2} + c_{1m} \ln r\right)\right\}\right), \quad (8.206)
\end{aligned}
$$

$$
\begin{aligned}
\sigma'_{12m} = \frac{1}{s_{44m}\zeta_m} &\left(\left(\frac{4}{3} - \ln r\right) \left[\zeta_m^{(11)} \frac{\partial p_p}{\partial \varphi} + p_p \frac{\partial \zeta_m^{(11)}}{\partial \varphi} - \frac{p_p \zeta_m^{(11)}}{\zeta_m} \frac{\partial \zeta_m^{(11)}}{\partial \varphi}\right] \right. \\
&+ r^{c_{3m}-1} \left[\zeta_m^{(21)} \frac{\partial p_p}{\partial \varphi} + p_p \frac{\partial \zeta_m^{(21)}}{\partial \varphi} - \frac{p_p \zeta_m^{(21)}}{\zeta_m} \frac{\partial \zeta_m^{(21)}}{\partial \varphi}\right] \\
&+ \frac{1}{r}\left(\frac{1}{2} + \ln r\right) \left[\zeta_m^{(31)} \frac{\partial p_p}{\partial \varphi} + p_p \frac{\partial \zeta_m^{(31)}}{\partial \varphi} - \frac{p_p \zeta_m^{(31)}}{\zeta_m} \frac{\partial \zeta_m^{(31)}}{\partial \varphi}\right] \\
&+ \frac{u_{1c0}}{d} \left\{ \left(\frac{4}{3} - \ln r\right) \left[\frac{\partial \zeta_m^{(12)}}{\partial \varphi} - \frac{\zeta_m^{(12)}}{\zeta_m} \frac{\partial \zeta_m}{\partial \varphi}\right] \right. \\
&\quad + r^{c_{3m}-1} \left[\frac{\partial \zeta_m^{(22)}}{\partial \varphi} - \frac{\zeta_m^{(22)}}{\zeta_m} \frac{\partial \zeta_m}{\partial \varphi}\right] \\
&\quad \left.\left. + \frac{1}{r}\left(\frac{1}{2} + \ln r\right) \left[\frac{\partial \zeta_m^{(32)}}{\partial \varphi} - \frac{\zeta_m^{(32)}}{\zeta_m} \frac{\partial \zeta_m}{\partial \varphi}\right]\right\}\right),
\end{aligned}
$$

(8.207)

$$
\begin{aligned}
\sigma'_{13m} = \frac{1}{s_{44m}\zeta_m} &\left(\left(\frac{4}{3} - \ln r\right) \left[\zeta_m^{(11)} \frac{\partial p_p}{\partial \nu} + p_p \frac{\partial \zeta_m^{(11)}}{\partial \nu} - \frac{p_p \zeta_m^{(11)}}{\zeta_m} \frac{\partial \zeta_m^{(11)}}{\partial \nu}\right] \right. \\
&+ r^{c_{3m}-1} \left[\zeta_m^{(21)} \frac{\partial p_p}{\partial \nu} + p_p \frac{\partial \zeta_m^{(21)}}{\partial \nu} - \frac{p_p \zeta_m^{(21)}}{\zeta_m} \frac{\partial \zeta_m^{(21)}}{\partial \nu}\right]
\end{aligned}
$$

$$+\frac{1}{r}\left(\frac{1}{2}+\ln r\right)\left[\zeta_m^{(31)}\frac{\partial p_p}{\partial \nu}+p_p\frac{\partial \zeta_m^{(31)}}{\partial \nu}-\frac{p_p\,\zeta_m^{(31)}}{\zeta_m}\frac{\partial \zeta_m^{(31)}}{\partial \nu}\right]$$

$$+\frac{u_{1c0}}{d}\left\{\left(\frac{4}{3}-\ln r\right)\left[\frac{\partial \zeta_m^{(12)}}{\partial \nu}-\frac{\zeta_m^{(12)}}{\zeta_m}\frac{\partial \zeta_m}{\partial \nu}\right]\right.$$

$$+r^{c_{3m}-1}\left[\frac{\partial \zeta_m^{(22)}}{\partial \nu}-\frac{\zeta_m^{(22)}}{\zeta_m}\frac{\partial \zeta_m}{\partial \nu}\right]$$

$$\left.\left.+\frac{1}{r}\left(\frac{1}{2}+\ln r\right)\left[\frac{\partial \zeta_m^{(32)}}{\partial \nu}-\frac{\zeta_m^{(32)}}{\zeta_m}\frac{\partial \zeta_m}{\partial \nu}\right]\right\}\right), \tag{8.208}$$

where s_{44m} is given by Equation (3.35).

8.5.2. Cell Matrix around Void

As presented in Section 5.3, the absolute values $|u'_{1v}|$, $\left|\varepsilon'_{ijv}\right|$, $\left|\sigma'_{ijv}\right|$ $(i,j=1,2,3)$ are required to represent increasing functions of $r \in \langle R_v, r_c \rangle$ (see Equations (3.19)–(3.23), (3.36)–(3.39)). These increasing functions exhibit maximum values on the cell surface, i.e., for $r = r_c$ (see Equations (3.15)–(3.18)). These increasing courses are ensured by the integration constants $C_{1v} \neq 0$, $C_{2v} \neq 0$, $C_{3v} \neq 0$, which are determined by the mandatory and additional boundary conditions (5.5), (5.6) and (5.8), respectively. With regard to Equations (5.5), (5.6), (5.8), (8.4)–(8.7), (8.15), we get

$$C_{iv} = -\frac{u'_{1c}\chi_{iv}}{\xi_{iv}\chi_v r_c} = -\frac{u_{1c0}\chi_{iv}}{d\,\xi_{iv}\chi_v}, \quad i=1,2,3, \tag{8.209}$$

where $u'_{1c}/r_c = u_{1c0}/d$ (see Equations (3.15)–(3.18), (5.4)); and r_c, ξ_{iv} $(i=1,2,3)$ are given by Equations (3.15)–(3.18), (7.8)–(7.10), respectively. The displacement u_{1c0} along the axis x_i $(i=1,2,3)$ on the cell surface *I*, *II* (see Equation (5.4)), i.e., for $r = d/2$ (see Figure 2.1), is determined in Section 7.5.4. The coefficients χ_{iv} $(i=1,2,3)$, χ_v, have the forms

$$\chi_{1v} = r_c\left\{c_{3m}\left(c_{1m}-2c_{2m}\ln R_v\right)\frac{\left(r_c-u'_{1c}\right)^{c_{3m}-1}}{R_v}\right.$$

$$- \left[(c_{1m} + c_{2m})\, c_{3m} - 2\, c_{2m} \right] \frac{R_v^{c_{3m}-1}}{r_c - u'_{1c}} \Bigg\},$$

(8.210)

$$\chi_{2v} = r_c \left\{ \frac{1}{r_c - u'_{1c}} \left[\frac{c_{1m} - 7\, c_{2m}}{3} - (c_{1m} - c_{2m}) \ln R_v \right] - \frac{c_{1m} - 2\, c_{2m} \ln R_v}{R_v} \left[\frac{1}{3} - \ln \left(r_c - u'_{1c} \right) \right] \right\} \tag{8.211}$$

$$\chi_{3v} = r_c \left\{ \left[(c_{1m} + c_{2m})\, c_{3m} - 2\, c_{2m} \right] \left[\frac{1}{3} - \ln \left(r_c - u'_{1c} \right) \right] R_v^{c_{3m}-1} - c_{3m} \left[\frac{c_{1m} - 7\, c_{2m}}{3} - (c_{1m} - c_{2m}) \ln R_v \right] \left(r_c - u'_{1c} \right)^{c_{3m}-1} \right\}, \tag{8.212}$$

$$\begin{aligned} \chi_v = {} & \frac{r_c^{c_{3m}}}{r_c - u'_{1c}} \left[\frac{c_{1m} - 7\, c_{2m}}{3} - (c_{1m} - c_{2m}) \ln R_v \right] \\ & + \left[(c_{1m} + c_{2m})\, c_{3m} - 2\, c_{2m} \right] \left(\frac{1}{2} + \ln r_c \right) \left[\frac{1}{3} - \ln \left(r_c - u'_{1c} \right) \right] R_v^{c_{3m}-1} \\ & + \left(c_{1m} - 2\, c_{2m} \ln R_v \right) \left(\frac{4}{3} - \ln r_c \right) \frac{r_c\, c_{3m} \left(r_c - u'_{1c} \right)^{c_{3m}-1}}{R_v} \\ & - \left\{ c_{3m} \left[\frac{c_{1m} - 7\, c_{2m}}{3} - (c_{1m} - c_{2m}) \ln R_v \right] \left(\frac{1}{2} + \ln r_c \right) \left(r_c - u'_{1c} \right)^{c_{3m}-1} \right. \\ & + \left[(c_{1m} + c_{2m})\, c_{3m} - 2\, c_{2m} \right] \left(\frac{4}{3} - \ln r_c \right) \frac{r_c R_v^{c_{3m}-1}}{r_c - u'_{1c}} \\ & \left. + \frac{\left(c_{1m} - 2\, c_{2m} \ln R_v \right) r_c^{c_{3m}}}{R_v} \left[\frac{1}{3} - \ln \left(r_c - u'_{1c} \right) \right] \right\}, \end{aligned} \tag{8.213}$$

where c_{im} $(i = 1,2,3)$ is given by Equation (3.40). With regard to Equations (7.4)–(7.7), (8.11)–(8.18), (8.209), we get

$$u'_{1v} = - \frac{u_{1c0}}{d\chi_v} \left[\chi_{1v}\, r \left(\frac{4}{3} - \ln r \right) + \chi_{2v}\, r^{c_{3m}} + \chi_{3v} \left(\frac{1}{2} + \ln r \right) \right], \tag{8.214}$$

$$\varepsilon'_{11v} = -\frac{u_{1c0}}{d\chi_v}\left[\chi_{1v}\left(\frac{1}{3} - \ln r\right) + \chi_{2v} c_{3m} r^{c_{3m}-1}\right], \tag{8.215}$$

$$\varepsilon'_{22v} = \varepsilon'_{33v} = -\frac{u_{1c0}}{d\chi_v}\left[\chi_{1v}\left(\frac{4}{3} - \ln r\right) + \chi_{2v} r^{c_{3m}-1} + \frac{\chi_{3v}}{r}\left(\frac{1}{2} + \ln r\right)\right], \tag{8.216}$$

$$\varepsilon'_{12v} = -\frac{u_{1c0}}{d\chi_v}\left\{\left(\frac{4}{3} - \ln r\right)\left[\frac{\partial\chi_{1v}}{\partial\varphi} - \frac{\chi_{1v}}{\chi_v}\frac{\partial\chi_v}{\partial\varphi}\right] + r^{c_{3m}-1}\left[\frac{\partial\chi_{2v}}{\partial\varphi} - \frac{\chi_{2v}}{\chi_v}\frac{\partial\chi_v}{\partial\varphi}\right] + \frac{1}{r}\left(\frac{1}{2} + \ln r\right)\left[\frac{\partial\chi_{3v}}{\partial\varphi} - \frac{\chi_{3v}}{\chi_v}\frac{\partial\chi_v}{\partial\varphi}\right]\right\}, \tag{8.217}$$

$$\varepsilon'_{13v} = -\frac{u_{1c0}}{d\chi_v}\left\{\left(\frac{4}{3} - \ln r\right)\left[\frac{\partial\chi_{1v}}{\partial\nu} - \frac{\chi_{1v}}{\chi_v}\frac{\partial\chi_v}{\partial\nu}\right] + r^{c_{3m}-1}\left[\frac{\partial\chi_{2v}}{\partial\nu} - \frac{\chi_{2v}}{\chi_v}\frac{\partial\chi_v}{\partial\nu}\right] + \frac{1}{r}\left(\frac{1}{2} + \ln r\right)\left[\frac{\partial\chi_{3v}}{\partial\nu} - \frac{\chi_{3v}}{\chi_v}\frac{\partial\chi_v}{\partial\nu}\right]\right\}, \tag{8.218}$$

$$\sigma'_{11v} = -\frac{u_{1c0}}{d\chi_v}\left\{\chi_{1v}\left[\frac{c_{1m} - 7c_{2m}}{3} - (c_{1m} - c_{2m})\ln r\right] + \chi_{2v}\left[(c_{1m} + c_{2m})c_{3m} - 2c_{2m}\right] r^{c_{3m}-1} + \frac{\chi_{3v}(c_{1m} - 2c_{2m}\ln r)}{r}\right\}, \tag{8.219}$$

$$\sigma'_{22v} = \sigma'_{33v} = -\frac{u_{1c0}}{d\chi_v}\left\{\zeta_m^{(11)}\left[\frac{4c_{1m} - c_{2m}}{3} - (c_{1m} - c_{2m})\ln r\right] + \zeta_m^{(21)}(c_{1m} - c_{2m}c_{3m}) r^{c_{3m}-1} + \zeta_m^{(31)}\left(\frac{c_{1m} - 2c_{2m}}{2} + c_{1m}\ln r\right)\right\}, \tag{8.220}$$

$$\sigma'_{12v} = -\frac{u_{1c0}}{d s_{44m}\chi_v}\left\{\left(\frac{4}{3} - \ln r\right)\left[\frac{\partial\chi_{1v}}{\partial\varphi} - \frac{\chi_{1v}}{\chi_v}\frac{\partial\chi_v}{\partial\varphi}\right]\right.$$

$$+ r^{c_{3m}-1}\left[\frac{\partial \chi_{2v}}{\partial \varphi} - \frac{\chi_{2v}}{\chi_v}\frac{\partial \chi_v}{\partial \varphi}\right]$$

$$+ \frac{1}{r}\left(\frac{1}{2} + \ln r\right)\left[\frac{\partial \chi_{3v}}{\partial \varphi} - \frac{\chi_{3v}}{\chi_v}\frac{\partial \chi_v}{\partial \varphi}\right]\Bigg\}, \quad (8.221)$$

$$\sigma'_{13v} = -\frac{u_{1c0}}{d\, s_{44m}\chi_v}\Bigg\{\left(\frac{4}{3} - \ln r\right)\left[\frac{\partial \chi_{1v}}{\partial \nu} - \frac{\chi_{1v}}{\chi_v}\frac{\partial \chi_v}{\partial \nu}\right]$$

$$+ r^{c_{3m}-1}\left[\frac{\partial \chi_{2v}}{\partial \nu} - \frac{\chi_{2v}}{\chi_v}\frac{\partial \chi_v}{\partial \nu}\right]$$

$$+ \frac{1}{r}\left(\frac{1}{2} + \ln r\right)\left[\frac{\partial \chi_{3v}}{\partial \nu} - \frac{\chi_{3v}}{\chi_v}\frac{\partial \chi_v}{\partial \nu}\right]\Bigg\}, \quad (8.222)$$

where s_{44m} is given by Equation (3.35).

8.5.3. Radial Stress p_p

With regard to Equations (5.12), (6.25), (8.202), the radial stress p_p, which acts at the particle-matrix boundary, i.e., for $r = R_p$ (see Figure 2.1a), has the form

$$p_p = \zeta_{pm}\left(\frac{u_{1c0}\zeta_{2m}}{d} + \beta_m - \beta_p\right), \quad (8.223)$$

where $u'_{1c}/r_c = u_{1c0}/d$ (see Equations (3.15)–(3.18), (5.4)). The coefficients ζ_{pm}, ζ_{1m}, ζ_{2m} are derived as

$$\zeta_{pm} = \frac{\xi_{3p}}{\xi_{3p}\zeta_{1m} - 1}, \quad (8.224)$$

$$\zeta_{1m} = \frac{1}{\zeta_m}\left[\zeta_m^{(11)}\left(\frac{4}{3} - \ln R_p\right) + \zeta_m^{(21)} R_p^{c_{3m}-1} + \frac{\zeta_m^{(31)}}{R_p}\left(\frac{1}{2} + \ln R_p\right)\right], \quad (8.225)$$

$$\zeta_{2m} = \frac{1}{\zeta_m}\left[\zeta_m^{(12)}\left(\frac{4}{3} - \ln R_p\right) + \zeta_m^{(22)} R_p^{c_{3m}-1} + \frac{\zeta_m^{(32)}}{R_p}\left(\frac{1}{2} + \ln R_p\right)\right], \quad (8.226)$$

where c_{3m}, ξ_{3p}, ζ_m, $\zeta_m^{(i1)}$, $\zeta_m^{(i2)}$ $(i = 1,2,3)$ are given by Equations (3.40), (6.19), (8.193)–(8.199), respectively.

8.5.4. Displacement u_{1c0}

With regard to Equations (8.192), (8.209), (8.223), the integration constant C_{iq} ($i = 1,2,3$; $q = m,v$) has the form

$$C_{im} = \frac{1}{\xi_{im}} \left[\frac{u_{1c0}\,\zeta_{i1m}}{d} + \zeta_{i2m}\left(\beta_m - \beta_p\right) \right], \quad i = 1,2,3, \tag{8.227}$$

$$C_{iv} = \frac{u_{1c0}\,\zeta_{iv}}{d\,\xi_{iv}}, \quad i = 1,2,3, \tag{8.228}$$

where ξ_{im}, ξ_{im} ($i = 1,2,3$) are given by Equations (7.8)–(7.10), and β_q ($q = p,m$) is determined in Section 4.2. The coefficients ζ_{i1m}, ζ_{i2m}, ζ_{iv} ($i = 1,2,3$) are derived as

$$\zeta_{i1m} = \frac{1}{\zeta_m} \left[\zeta_m^{(i1)}\,\zeta_{pm}\,\zeta_{2m} + \zeta_m^{(i2)} \right], \quad i = 1,2,3, \tag{8.229}$$

$$\zeta_{i2m} = \frac{\zeta_m^{(i1)}\,\zeta_{pm}}{\zeta_m}, \quad i = 1,2,3, \tag{8.230}$$

$$\zeta_{iv} = -\frac{\chi_{iv}}{\chi_v}, \quad i = 1,2,3, \tag{8.231}$$

where ζ_m, $\zeta_m^{(i1)}$, $\zeta_m^{(i2)}$, χ_v, χ_{iv} ($i = 1,2,3$), ζ_{pm}, ζ_{2m} are given by Equations (8.193)–(8.199), (8.210)–(8.212), (8.224), (7.226), respectively.

The elastic energy W_m, which is accumulated in the cell matrix *I*, is given by Equation (8.78). The coefficient $\Omega_{im}^{(12)}$ ($i = 1,2,3$), which is included in Equation (8.78), is replaced by $\Omega_{im}^{(123)} = \Omega_{im}^{(12)} + \Omega_{im}^{(13)} + \Omega_{im}^{(23)}$, where $\Omega_{im}^{(12)}$, $\Omega_{im}^{(13)}$, $\Omega_{im}^{(23)}$ are given by Equations (8.79)–(8.81), (8.136)–(8.138), (8.186)–(8.188), respectively. The coefficients ζ_{i1m}, ζ_{i2m} ($i = 1,2,3$), which are included in Equations (8.79)–(8.81), (8.136)–(8.138), (8.186)–(8.188), are given by Equations (8.225), (8.226), respectively.

The elastic energy W_v, which is accumulated in the cell matrix *II*, is given by Equation (8.89). The coefficient $\Omega_v^{(12)}$ ($i = 1,2,3$), which is included in Equation (8.89), is replaced by $\Omega_v^{(123)} = \Omega_{imv}^{(12)} + \Omega_v^{(13)} + \Omega_v^{(23)}$, where $\Omega_v^{(12)}$, $\Omega_v^{(13)}$, $\Omega_v^{(23)}$ are given by Equations (8.90), (8.143), (8.191), respectively. The coefficient ζ_{iv} ($i = 1,2,3$), which is included in Equations (8.90), (8.143), (8.191), is given by Equation (8.231).

With regard to Equations (8.183)–(8.199), (8.210)–(8.213), the elastic energy $W = W_p + W_m - W_v$ (see Equation (5.13)) is not a polynomial function of the variable u_{1c0}. The condition (5.13) determines a minimum of the function $W = W(u_{1c0})$. This minimum is determined for a real two-component material by a numerical method. Finally, such a numerical value of the variable u_{1c0} is determined, which results in a minimum numerical value of $W = W_p + W_m - W_v$.

Chapter 9

Selected Topics in Mathematics

This chapter presents such a part of mathematics, which is required within the mathematical procedures in this book.

9.1. Cramer's Rule

In case of $n \geq 3$, the determinant $D^{(n)}$ with n rows and n columns has the form [25]

$$D^{(n)} = \begin{vmatrix} a_{11}, & a_{12}, & \ldots & a_{1n} \\ a_{21}, & a_{22}, & \ldots & a_{2n} \\ \vdots & \vdots & \vdots & \vdots \\ a_{n1}, & a_{n2}, & \ldots & a_{nn} \end{vmatrix}$$

$$= \sum_{i=1}^{n} (-1)^{1+i} a_{1i} D_{1i}^{(n-1)} = \sum_{i=1}^{n} (-1)^{1+i} a_{i1} D_{i1}^{(n-1)}, \tag{9.1}$$

where a_{ij} $(i, j = 1, \ldots, \mathrm{n})$ is a coefficient. The subdeterminant $D_{ij}^{(n-1)}$ $(i, j = 1, \ldots, \mathrm{n})$ with $(n-1)$ rows and $(n-1)$ columns is created from $D^{(n)}$, where the i-th row and the j-th column j of the determinant $D^{(n)}$ are omitted. In case of $n = 2$, the determinant $D^{(2)}$ is derived as

$$D^{(2)} = \begin{vmatrix} a_{11}, & a_{12} \\ a_{21}, & a_{22} \end{vmatrix} = a_{11}a_{22} - a_{12}a_{21}. \tag{9.2}$$

In case of $n = 4$, we get

$$
\begin{aligned}
D^{(4)} &= \begin{vmatrix} a_{11}, & a_{12}, & a_{13}, & a_{14} \\ a_{21}, & a_{22}, & a_{23}, & a_{24} \\ a_{31}, & a_{32}, & a_{33}, & a_{34} \\ a_{41}, & a_{42}, & a_{43}, & a_{44} \end{vmatrix} \\
&= a_{11} \begin{vmatrix} a_{22}, & a_{23}, & a_{24} \\ a_{32}, & a_{33}, & a_{34} \\ a_{42}, & a_{43}, & a_{44} \end{vmatrix} - a_{12} \begin{vmatrix} a_{21}, & a_{23}, & a_{24} \\ a_{31}, & a_{33}, & a_{34} \\ a_{41}, & a_{43}, & a_{44} \end{vmatrix} \\
&\quad + a_{13} \begin{vmatrix} a_{21}, & a_{22}, & a_{24} \\ a_{31}, & a_{32}, & a_{34} \\ a_{41}, & a_{42}, & a_{44} \end{vmatrix} - a_{14} \begin{vmatrix} a_{21}, & a_{22}, & a_{23} \\ a_{31}, & a_{32}, & a_{33} \\ a_{41}, & a_{42}, & a_{43} \end{vmatrix} \\
&= a_{11} \left(a_{22} \begin{vmatrix} a_{33}, & a_{34} \\ a_{43}, & a_{44} \end{vmatrix} - a_{23} \begin{vmatrix} a_{32}, & a_{34} \\ a_{42}, & a_{44} \end{vmatrix} + a_{24} \begin{vmatrix} a_{32}, & a_{33} \\ a_{42}, & a_{43} \end{vmatrix} \right) \\
&\quad - a_{12} \left(a_{21} \begin{vmatrix} a_{33}, & a_{34} \\ a_{43}, & a_{44} \end{vmatrix} - a_{23} \begin{vmatrix} a_{31}, & a_{34} \\ a_{41}, & a_{44} \end{vmatrix} + a_{24} \begin{vmatrix} a_{31}, & a_{33} \\ a_{41}, & a_{43} \end{vmatrix} \right) \\
&\quad + a_{13} \left(a_{21} \begin{vmatrix} a_{32}, & a_{34} \\ a_{42}, & a_{44} \end{vmatrix} - a_{22} \begin{vmatrix} a_{31}, & a_{34} \\ a_{41}, & a_{44} \end{vmatrix} + a_{24} \begin{vmatrix} a_{31}, & a_{32} \\ a_{41}, & a_{42} \end{vmatrix} \right) \\
&\quad - a_{14} \left(a_{21} \begin{vmatrix} a_{32}, & a_{33} \\ a_{42}, & a_{43} \end{vmatrix} - a_{22} \begin{vmatrix} a_{31}, & a_{33} \\ a_{41}, & a_{43} \end{vmatrix} + a_{23} \begin{vmatrix} a_{31}, & a_{32} \\ a_{41}, & a_{42} \end{vmatrix} \right) \\
&= (a_{11}a_{22} - a_{12}a_{21})(a_{33}a_{44} - a_{34}a_{43}) + (a_{14}a_{22} - a_{12}a_{24})(a_{31}a_{43} - a_{33}a_{41}) \\
&\quad + (a_{13}a_{24} - a_{14}a_{23})(a_{31}a_{42} - a_{32}a_{41}) + (a_{12}a_{23} - a_{13}a_{22})(a_{31}a_{44} - a_{34}a_{41}) \\
&\quad + (a_{11}a_{24} - a_{14}a_{21})(a_{32}a_{43} - a_{33}a_{42}) + (a_{13}a_{21} - a_{11}a_{23})(a_{32}a_{44} - a_{34}a_{42})
\end{aligned} \tag{9.3}
$$

Let $x_1, \ldots, x_n$ be roots of a system of n linear algebraic equations, which have the forms [25]

$$
\begin{aligned}
a_{11}x_1 + a_{12}x_2 + \ldots + a_{1n}x_n &= b_1, \\
a_{11}x_1 + a_{12}x_2 + \ldots + a_{1n}x_n &= b_2, \\
\vdots \qquad \vdots \qquad \vdots \qquad &\vdots \\
a_{n1}x_1 + a_{n2}x_2 + \ldots + a_{nn}x_n &= b_n.
\end{aligned} \tag{9.4}
$$

The root x_i ($i = 1, \ldots,$ n) is determined by Cramer's rule, which is derived as [25]

$$x_i = \frac{D_i^{(n)}}{D^{(n)}}, \quad i = 1, \ldots, n, \tag{9.5}$$

where the determinant $D^{(n)}$ is given by Equation (9.1). The determinant $D_i^{(n)}$ is created from $D^{(n)}$, where the i-th column of the determinant $D^{(n)}$ is replaced by the following column

$$\left.\begin{matrix} b_1 \\ b_2 \\ \vdots \\ b_n \end{matrix}\right\} n \text{ rows}. \tag{9.6}$$

9.2. Quadratic Algebraic Equation

Let λ_i, λ_2 be roots of the following quadratic algebraic equation [25]

$$a_1\lambda^2 + a_2\lambda + a_3 = 0, \tag{9.7}$$

where $a_1 \neq 0$, a_2, a_3 are coefficients. The roots λ_i, λ_2 and the discriminant D have the forms [25]

$$\lambda_{1,2} = \frac{-a_2 \pm \sqrt{D}}{2a_1}, \tag{9.8}$$

$$D = a_2^2 - 4a_1a_3. \tag{9.9}$$

If $D < 0$, then $\sqrt{D}$ is replaced by $i\sqrt{D}$, where $i = \sqrt{-1}$ is an imaginary unit [25].

9.3. Determination of Integrals

The elastic energy W is determined by indefinite and definite integrals (see Sections 6.6, 7.2.4, 7.3.4, 7.4.4, 8.2.4, 8.3.4, 8.4.4). Consequently, the following analysis is considered within this determination.

Derivatives of the functions $f = r^\lambda$, $f = \ln r$, $C \neq f(r)$, which are applied in this book, are as follows [25]

$$\left(r^{\lambda}\right)' = \lambda r^{\lambda-1}, \tag{9.10}$$

$$(\ln r)' = \frac{1}{r}. \tag{9.11}$$

$$C' = 0, \tag{9.12}$$

where $C \neq f(r)$ is a constant function with respect to the variable r. Similarly, indefinite integrals of $f = r^{\lambda}$, $f = \ln r$, $C \neq f(r)$ have the forms [25]

$$\int r^{\lambda} dr = \frac{r^{\lambda}+1}{\lambda+1}, \quad \lambda \neq -1, \tag{9.13}$$

$$\int \frac{dr}{r} = \ln r, \tag{9.14}$$

$$\int C dr = Cr. \tag{9.15}$$

A derivation of the product fg of the functions $f = f(r)$, $g = g(r)$ with the variable r is derived as [25]

$$(fg)' = f'g + fg'. \tag{9.16}$$

Consequently, the following formula [25]

$$\int f'g\,dr = fg - \int fg'\,dr. \tag{9.17}$$

is used to determine indefinite integrals. In case of $f' = r^{\lambda}$, $g = \ln r$, the indefinite integral $\int r^{\lambda} \ln r\, dr$ for $\lambda \neq -1$ is determined by the following mathematical procedure (see Equation (9.17))

$$\int r^{\lambda} \ln r\, dr = \frac{r^{\lambda+1}}{\lambda+1} \ln r - \int \frac{r^{\lambda+1}}{\lambda+1} \times \frac{1}{r} dr = \frac{r^{\lambda+1}}{\lambda+1} \ln r - \frac{1}{\lambda+1} \int r^{\lambda} dr$$
$$= \frac{r^{\lambda+1}}{\lambda+1}\left(\ln r - \frac{1}{\lambda+1}\right), \quad \lambda \neq -1. \tag{9.18}$$

In case of $f' = 1$, $g = \ln r$, $C = 1$, the indefinite integral $\int \ln r\, dr$ is determined by the following mathematical procedure (see Equation (9.17))

$$\int \ln r\, dr = \int 1 \times \ln r\, dr = r \ln r - \int r \times \frac{1}{r}\, dr = r \ln r - \int 1 \times dr = r\left(\ln r - 1\right). \tag{9.19}$$

The indefinite integral $\int r^{\lambda} \ln^2 r\, dr$ for $\lambda \neq -1$ is determined by the following mathematical procedure (see Equation (9.17))

$$\begin{aligned} \int r^{\lambda} \ln^2 r\, dr &= \frac{1}{\lambda+1}\left[r^{\lambda+1} \ln^2 r - 2\int r^{\lambda} \ln r\, dr\right] \\ &= \frac{r^{\lambda+1}}{\lambda+1}\left[\left(\ln r - \frac{1}{\lambda+1}\right)^2 + \frac{1}{(\lambda+1)^2}\right], \quad \lambda \neq -1. \end{aligned} \tag{9.20}$$

Let $F = F(r)$ be a primitive function of $f = f(r)$ in the interval $r \in \langle a, b \rangle$, i.e., $f = dF/dr$. The definite integral $\int_a^b f\, dr$ is defined by the Newton-Leibniz's formula [25], which has the form

$$\int_a^b f\, dr = F(b) - F(a). \tag{9.21}$$

9.4. Wronskian's Method

This section presents an application of Wronskian's method to Equation (6.6) with a non-zero right-hand side, which is derived as

$$\frac{\partial^2 u_1'}{\partial r^2} + \frac{2}{r}\frac{\partial u_1'}{\partial r} - \frac{2u_1'}{r^2} = g. \tag{9.22}$$

The non-zero right-hand side has the form

$$g = -\frac{1}{s_{44}\left(c_1 + c_2\right)} \sum_{i=1}^{2} C_i r^{\lambda_i - 2}, \tag{9.23}$$

where s_{44} and λ_i $(i = 1,2)$ are given by Equations (3.35) and (6.4), respectively. The integration constant C_i $(i = 1,2)$ is determined by the mathematical boundary conditions, which are defined in Chapter 5. Equation (6.6) represents a linear differential equation of the second rank with respect to the variable r. If $g = 0$, we get

$$\frac{\partial^2 u_1'}{\partial r^2} + \frac{2}{r}\frac{\partial u_1'}{\partial r} - \frac{2u_1'}{r^2} = 0. \tag{9.24}$$

Let u_1' be assumed in the form $u_1' = r^\lambda$. With regard to $u_1' = r^\lambda$, the differential equation (9.23) is transformed to a following characteristic equation, which is derived as

$$\lambda^2 + \lambda - 2 = 0, \tag{9.25}$$

where $\lambda_1 = 1$ and $\lambda_2 = -2$ are roots of this characteristic equation (see Equations (9.7)–(9.9)). Accordingly, the functions

$$u_{11}' = r^{\lambda_1} = r, \quad u_{12}' = r^{\lambda_2} = \frac{1}{r^2}, \tag{9.26}$$

represent solutions of the differential equation (9.24) with a zero right-hand side, where $\partial u_{1i}'/\partial r$ $(i = 1,2)$ has the form

$$\frac{\partial u_{11}}{\partial r} = 1, \quad \frac{\partial u_{12}}{\partial r} = -\frac{2}{r^3}. \tag{9.27}$$

The solution u_1' of the differential equation (9.22) with a non-zero right-hand side (see Equation (9.23)) is derived as [25]

$$u_1' = \sum_{i=1}^{2} a_i u_{1i}' = \sum_{i=1}^{2} C_i \xi_i u_{ir}, \tag{9.28}$$

where u_{ir}, ξ_i $(i = 1,2)$ are given by Equations (6.8), (6.9), respectively, and a_i $(i = 1,2)$ has the form [25]

$$a_i = \int \frac{W_i^{(2)}}{W^{(2)}}\, dr. \tag{9.29}$$

The Wronskian's determinants $W^{(2)}$, $W_i^{(2)}$ $(i = 1,2)$ with 2 rows and 2 columns are derived as [25]

$$W^{(2)} = \begin{vmatrix} u_{11}, & u_{12} \\ \frac{\partial u_{11}}{\partial r}, & \frac{\partial u_{12}}{\partial r} \end{vmatrix}, \tag{9.30}$$

$$W_1^{(2)} = \begin{vmatrix} 0, & u_{12} \\ g, & \dfrac{\partial u_{12}}{\partial r} \end{vmatrix}, \tag{9.31}$$

$$W_2^{(2)} = \begin{vmatrix} u_{11}, & 0 \\ \dfrac{\partial u_{11}}{\partial r}, & g \end{vmatrix}. \tag{9.32}$$

The determinant $W_i^{(2)}$ $(i=1,2)$ is created from $W^{(2)}$, where the i-th column of the determinant $W^{(2)}$ is replaced by the following column [25]

$$\left.\begin{matrix} 0 \\ g \end{matrix}\right\} \text{2 rows.} \tag{9.33}$$

Let $u'_{11},\ldots,u'_{1n}$ represent solutions of a differential equation of the n-th rank with zero right-hand side, where $u'_{11},\ldots,u'_{1n}$ are functions of the variable r. Let these solutions exhibit continuous derivatives to the $(n-1)$-th degree with respect to the variable r. A solution of this differential equation with a non-zero right-hand side (i.e., $g \neq 0$) represents linear combination of the terms $a_{1q}u_{11}, \ldots, a_{nq}u_{1n}$ [25]. Consequently, we get [25]

$$u'_1 = \sum_{i=1}^{n} a_{iq}\, u'_{1i}, \tag{9.34}$$

where a_i $(i=1,\ldots,\text{n})$ is derived as [25]

$$a_i = \int \frac{W_i^{(n)}}{W^{(n)}}\, dr. \tag{9.35}$$

The Wronskian's determinant $W^{(n)}$ $(i=1,\ldots,\text{n})$ with n rows and n columns is derived as [25]

$$W^{(n)} = \begin{vmatrix} u_{11}, & u_{12}, & \cdots & u_{1n} \\ \dfrac{\partial u_{11}}{\partial r}, & \dfrac{\partial u_{12}}{\partial r}, & \cdots & \dfrac{\partial u_{1n}}{\partial r} \\ \vdots & \vdots & \vdots & \vdots \\ \dfrac{\partial^{n-1} u_{11}}{\partial r^{n-1}}, & \dfrac{\partial^{n-1} u_{12}}{\partial r^{n-1}}, & \cdots & \dfrac{\partial^{n-1} u_{1n}}{\partial r^{n-1}} \end{vmatrix}. \tag{9.36}$$

The determinant $W_i^{(n)}$ $(i = 1, \ldots, n)$ with n rows and n columns is created from $W^{(n)}$, where the i-th column of the determinant $W^{(n)}$ is replaced by the following column [25]

$$\left.\begin{matrix} 0 \\ 0 \\ \vdots \\ g \end{matrix}\right\} n \text{ rows.} \tag{9.37}$$

9.5. Numerical Determination

Numerical values of the thermal and phase-transformation-induced stresses in a real two-component material include integrals and derivatives, which are required to be determined by suitable programming techniques.

Let $\psi = \psi(\eta)$ represent a function of the variable $\eta = \varphi, \nu$. Numerical values of the derivative $\partial\psi/\partial\eta$ are determined by the following formula

$$\frac{\partial\psi}{\partial\eta} \approx \frac{\psi(\eta + \Delta\eta) - \psi(\eta)}{\Delta\eta}, \tag{9.38}$$

where the angle steps $\Delta\psi = \Delta\varphi = \Delta\nu = 10^{-6}$ [deg] are sufficient.

Numerical values of a definite integral of the function $\omega = \omega(\varphi, \nu)$ with the variables $\varphi, \nu \in \langle 0, \pi/2 \rangle$ are determined by the following formula

$$\Omega = \int_0^{\pi/2}\int_0^{\pi/2} \omega d\varphi d\nu \approx \sum_{j=0}^{m}\left(\sum_{i=0}^{n} \omega(i \times \Delta\varphi; j \times \Delta\nu)\Delta\varphi\right)\Delta\nu, \tag{9.39}$$

where n, m are integral parts of the real numbers $\pi/(2\Delta\varphi)$, $\pi/(2\Delta\nu)$, respectively. The angle steps $\Delta\varphi = \Delta\nu = 0.01 - 0.1$ [deg] are sufficient.

The average numerical value $\overline{p_p}$ of the radial stress $p_p = p_p(\varphi, \nu)$ for $\varphi, \nu \in \langle 0, \pi/2 \rangle$, which acts at the particle-matrix boundary, is determined by the following formula

$$\overline{p_p} = \left(\frac{2}{\pi}\right)^2 \int_0^{\pi/2}\int_0^{\pi/2} p_p d\varphi d\nu \approx \left(\frac{2}{\pi}\right)^2 \sum_{j=0}^{m}\left(\sum_{i=0}^{n} p_p(i \times \Delta\varphi; j \times \Delta\nu)\Delta\varphi\right)\Delta\nu. \tag{9.40}$$

References

[1] Kushch, V.I. *Int. Appl. Mech.* **2004,** *40*, 1042-1049.

[2] Kushch, V.I.; Shmegera, S.V.; Mishnaevsky, L. Jr. *Int. J. Sol. Struct.* **2008,** *45*, 5103-5117.

[3] Broberg, K.B. *Comput. Mech.* **1997,** *19*, 447–452.

[4] Babuška, I.; Andersson, B.; Smith, P.J.; Levin, L.; *Computer Methods in Appl. Mech. Eng.* **1999,** *172*, 27–77.

[5] Mura, T. *Micromechanics of Defects in Solids;* Martinus Nijhoff Publishers: Dordrecht, NL, 1987; pp 388–392.

[6] Brdička, M.; Samek, L.; Sopko, B. *Mechanics of Continuum;* Academia: Prague, CZ, 2000; pp 205–206.

[7] Eshelby, J.D. *Proc. Royal Soc. London A* **1957,** *241*, 376-396.

[8] Li, S.; Sauer, R.A.; Wang, G. *J. Appl. Mech.* **2007,** *74*, 770–783.

[9] Li, S.; Sauer, R.A.; Wang, G. *J. Appl. Mech.* **2007,** *74*, 784–797.

[10] Trebuňa, F.; Šimčák, F.; Buršák, M.; Bocko, J.; Šarga, P.; Pástor, M.; Trebuňa, P. *Metalurgija* **2007,** *46*, 41–46.

[11] Shin, H.; Earmme, Y.Y. *Int. J. Fract.* **2004,** *126*, L35–L40.

[12] Hajko, V.; Potocký, L.; Zentko, A. *Magnetization Processes;* Alfa: Bratislava, SK, 1982; pp 37–45.

[13] Davidge, R.W.; Green, T.J. *J. Mater. Sci.* **1968,** *3*, 629–634.

[14] Diko, P. *Supercond. Sci. Technol.* **2004,** *17*, R45–R58.

[15] Chmelík, F.; Trník, A.; Štubňa, I.;, Pešička, J. *J. Eur. Ceram. Soc.* **2011,** *31*, 2205-2209.

[16] Selsing, J. *J. Amer. Cer. Soc.* **1961,** *44*, 419–419.

[17] Mizutani, T. *J. Mater. Sci.* 1998, *11*, 68–72.

[18] Skočovský, P.; Bokůvka, O.; Palček, P. *Materials Science*; EDIS: Žilina, SK, 1996; pp 75–79.

[19] Bidulský, R.; Bidulská, J.; Grande, M.A. *Metal Sci. Heat Treat.* **2017,** *58,* 734–737.

[20] Bidulský, R.; Bidulská, J.; Grande, M.A. *Arch. Metallur. Mater.* **2013,** *58,* 365–370.

[21] Kottfer, D.; Maňková, I.; Vrabeĺ M.; Kianicová, M.; Rehák, F.; Franková, M. *Sol. State Phenom.* **2017,** *261,* 237–242.

[22] Kottfer, D.; Ferdinandy, M.; Kaczmarek, L.; Trebuňa, P.; Hvizdoš, P. *Arch. Metallur. Mater.* **2016,** *61,* 67–73.

[23] Kianicová, M.; Kafrík, J. *Sol. State Phenom.* **2017,** *226,* 177–182.

[24] Kianicová, M.; Pokluda, J. *Key Engin. Mater.* **2014,** *592-593,* 712–715.

[25] Rektorys, K. *Review of Applied Mathematics;* SNTL: Prague, CZ, 1973; pp 156–162.

About the Author

Dr. Ladislav Ceniga, DSc was born in 1965 in Košice, Slovak Republic, graduated from the Faculty of Mechanical Engineering, the Technical University in Košice (1988) (Department of Mechanical Engineering Technology), and from the Faculty of Sciences of the P. J. Šafárik University in Košice (1993) (Department of Physics of Solids), both with distinction, awarded the prize of the Chancellor of the Technical University in Košice for excellent study results, aimed at heat and chemical treatment and magnetic properties of amorphous alloys (1994–2000), defending a PhD thesis in Physics of Condensed Matters and Acoustics at the Institute of Experimental Physics of the Slovak Academy of Sciences in Košice (1999), defending a DSc (Doctor of Sciences) thesis before the commission 'Mechanics of Bodies, Structures, Mechanisms and Environment' (2013) at the Institute of Thermomechanics of the Academy of Sciences of the Czech Republic in Prague, since 2000 employed at the Institute of Materials Research of the Slovak Academy of Sciences in Košice, Dr. Ceniga, DSc works on analytical models of thermal, phase-transformation-induced, hydrogen-induced stresses and related stress-induced phenomena in composite materials.

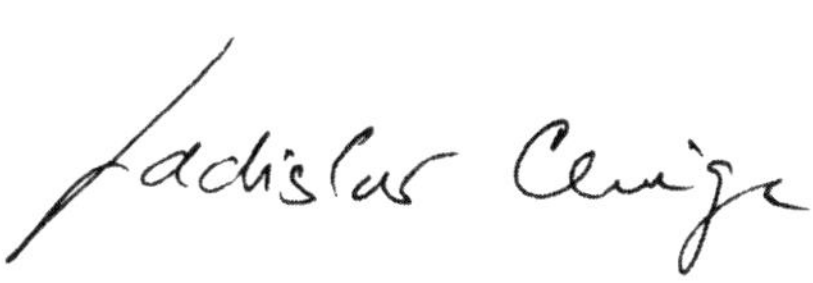

Dr. Ladislav Ceniga, DSc. Košice, Slovak Republic, 2018.

Address correspondence to: lceniga@yahoo.com; ladislav.ceniga@yahoo.com; ladislavceniga65@gmail.com; lceniga@hotmail.com

Index